The Biochemistry and Pharmacology of Antibacterial Agents

R.A.D. WILLIAMS and Z.L. KRUK

CROOM HELM LONDON

British Library Cataloguing in Publication Data

Williams, R.A.D.
    The biochemistry and pharmacology of antibacterial
    agents — (Croom Helm biology in medicine series)
    1. Antibacterial agents
    I. Title   II. Kruk, Zygmunt L
    615'.1   RM409

    ISBN 0-85664-858-2

Printed in the USA

THE BIOCHEMISTRY AND PHARMACOLOGY OF
ANTIBACTERIAL AGENTS

CROOM HELM BIOLOGY IN MEDICINE SERIES

STEROID HORMONES
D.B. Gower

NEUROTRANSMITTERS AND DRUGS
Zygmunt L. Kruk and Christopher J. Pycock

DEVELOPMENT, GROWTH AND AGEING
Edited by Nicholas Carter

INBORN ERRORS OF METABOLISM
Edited by Roland Ellis

MEMBRANE PHYSIOLOGY AND CELL EXCITATION
Bruce Hendry

CONTENTS

Contents

FOREWORD

Antibiotics are amongst the most widely used of drugs and the effectiveness of the multitude of agents available in curing fatal and severely disabling illness places them in the highest category of pharmaceuticals in terms of benefit to man. Infections, unlike many of the other major categories of diseases, can attack the person who is otherwise perfectly normal and curing of the infection can restore the patient to his original state of health. Many of the early studies on penicillin lay emphasis on its almost miraculous effects on previously fatal diseases.

The wide range of infecting agents in man determines the need for a wide range of substances with which to combat the infections, agents which can attack the differing vulnerable sites on the microbial cells. A much clearer concept of how to use antibiotics clinically is obtained if there is an understanding of the way the substances work, an understanding of what are the targets of antimicrobial action and how available are the targets of the different cells to the antibiotic. Without this basic knowledge, use of antibiotics clinically can become a speculative exercise in which we merely hope we have chosen the most appropriate of the array of substances for a particular patient. In this book the authors explain the ways in which antibiotics can disrupt the normal processes of bacterial growth, metabolism or replication, leading either to death of the microbial cell or to an inhibition of its processes to such an extent that the defence mechanisms of the host are able to overcome the invader. This small monograph should do much to provide the future prescriber of antibiotics and others whose work involves antibiotic use with a clear picture of the intricate mechanisms which are the focus of antibiotic action and the ways in which the antibiotics can alter the mechanisms. Antibiotics are so valuable it is impossible to learn too much about them.

J.D. Williams
Professor of Medical Microbiology
The London Hospital Medical College

PREFACE

Antimicrobial drugs are amongst the most important compounds in medicine and by their use infectious bacterial diseases have been largely abolished as a major cause of loss of life in more affluent countries. These substances are often dealt with in the traditional medical curriculum at various times by different specialists. Biochemists are concerned with the basic cellular processes against which the antimicrobial agents are effective, and often introduce the teaching of the subject by referring to the more important drugs when describing the metabolic process that they affect. Pharmacologists are concerned with the absorption, distribution, metabolic transformation and untoward effects of the antimicrobial agents, amongst many other drugs. Later in the curriculum medical microbiologists describe the types of micro-organisms against which the compounds are particularly effective, the testing of clinical samples for their sensitivities, and the important problems associated with resistant strains. In integrated curricula a more satisfactory unified approach to the subject may be possible, but one problem for the student that remains is that of availability of information, which is often scattered. The objective of this book is to unite the biochemical and pharmacological aspects of the use of common antimicrobials in a brief book suitable for the preclinical medical course, for medical microbiology units of science degrees and for some postgraduate courses.

1 INTRODUCTION

1.1 Historical Introduction

The earliest chemotherapeutic agents known to man were of plant origin. The Greeks used extract of male fern to treat infestation by worms, and extracts of cinchona bark were used in South America to treat malaria. Mercury, which was used to treat syphilis, was the first substance not of plant origin to be used chemotherapeutically. Until the beginning of the twentieth century preparations such as these were the only chemotherapeutic agents available.

At about this time, as an extension of his work on dyes and differential staining of tissues, Ehrlich developed the concept of cell surface *receptors*. These were defined as sites which were recognised by chemicals of a particular structure and to which the chemical became attached. Ehrlich considered that different cells might have different receptors to which chemicals could bind selectively. He reasoned that chemicals which bound to and were toxic towards pathogenic cells or organisms (but which did not bind to and therefore were not toxic to the host's cells) would be ideal drugs to treat infections. In the event, Ehrlich discovered arsphenamine (Salvarsan), an arsenical drug which was used to treat syphilis until it was superseded by penicillin in the 1940s. Work on dyes as chemotherapeutic agents continued, leading to the development of sulphonamides (see Chapter 2).

In 1928, Fleming observed that extracts from *Penicillium* could inhibit the growth of certain bacteria. Ten years later, Chain and Florey proved that such extracts could inhibit bacterial growth in mice as well as in cultures, and when in 1941 Abraham demonstrated the effectiveness of penicillin in man the era of antibiotic chemotherapy had begun. The search for other antibiotics was undertaken in many centres and to date about 5000 different substances have been discovered, many of which have found a use in chemotherapy of human infection.

1.2 Some Definitions

Chemotherapy is the study and practice of the use of drugs to damage an invading species without damage to the host. Since the time this definition was made by Ehrlich, the term chemotherapy has been applied not only to drug treatment of parasitic infestation by unicellular and multicellular organisms, but also to drug treatment of cancer. Whereas only those chemotherapeutic agents which are active against bacteria will be considered in this book, the principle of *selective toxicity* is the basis of all chemotherapy. The selective toxicity of chemotherapeutic agents aims to exploit differences in the biochemistry of host and parasite, or differences between normal and cancer cells, thus causing damage to the microorganisms or tumour cells while leaving the host's biochemistry undamaged.

Antimicrobial agents are those chemotherapeutic compounds which are toxic to microorganisms. Antimicrobial agents may be used outside the body, on the surface of the body, or systemically. *Sterilisation* is the destruction of all microbial life by chemical or physical means. Thus *disinfectants* such as phenols, 'Lysol' and sodium hypochlorite are caustic or toxic chemicals which are used in drains and places with heavy microbial populations. Heat and radiation are alternative methods of sterilisation. *Sanitising agents*, *antiseptics* or *germicides* (these terms are used interchangeably) do not sterilise completely, but they reduce the microbial population to acceptable levels. These substances are milder than disinfectants, may be suitable for use on living tissues and include, for example, dilute solutions of chlorinated phenols and acriflavine in wound dressings.

Antibiotics are chemicals produced by microorganisms (bacteria, fungi and actinomycetes) which are highly toxic to other microorganisms when the chemical is present in very low concentrations. Some antibiotics can be chemically synthesised, but most are made by culturing microorganisms under closely-controlled conditions. Whereas some antimicrobials are *bactericidal*, i.e. they kill the bacteria, in the treatment of disease it is frequently not necessary to kill all the infective microorganisms. A compound which is *bacteristatic*, i.e. it prevents further growth of the bacteria, allows the host's normal defence mechanisms to clear up the infection. The two types of action are not clearly distinguishable because the only test of whether bacteria are alive is their successful growth. It is important to remember that most chemotherapeutic agents *do not* eradicate all the infective organisms, but they allow the *body defences* to do this, and thus normal immunological and phagocytic mechanisms must be working efficiently if such drugs are to be effective.

The antibiotics (which comprise the majority of the antimicrobial drugs) are used in very large quantities today. Table 1.1 indicates the amounts of major antibiotics described in this book which were used by The London Hospital in 1978/9.

Table 1.1: Antibiotic Use at The London Hospital (1978/9)

	kg/year
Penicillins	121.0
Erythromycin	10.1
Cephalosporins	8.1
Aminoglycosides	2.1
Fusidate	1.4
Tetracyclines	1.35
Chloramphenicol	0.6

Note: This table provides an indication of the use of antibiotics in a large hospital. The therapeutic doses of the various antibiotics are quite different and the weights used are not an indication of the numbers of cases treated. We are grateful to the Drug Information Centre of The London Hospital Pharmacy for providing the data on which this table is based.

1.3 Sensitivity of Bacteria and the Development of Antibiotic Resistance

The factors that restrict the usefulness of a particular antibiotic include the differential sensitivity of the various infectious microorganisms, and the way in which this may change after exposure to the antibiotic. Intrinsic resistance may occur because the drug cannot gain access to its site of action in the bacteria. An important example of this is the failure of many penicillins to reach the peptidoglycan target in the cell wall of Gram-negative (i.e. do not stain purple with Grams stain) bacteria (see 3.2.1). These microorganisms have an outer hydrophobic cell-wall layer through which penicillin G cannot penetrate. Semi-

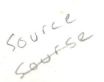

synthetic penicillin derivatives can pass through this outer-wall layer and are thus active against a broader spectrum of bacteria than the other members of the penicillin family.

Bacteria which are exposed to antibiotics to which they are sensitive are either killed or their growth is suppressed. Therefore spontaneous mutants that have any character which confers resistance will be selected by the drug and will proliferate. For this reason unnecessary use of antibiotics, which increases the exposure of bacterial populations to this selective force, is to be avoided.

One way of reducing the development of resistant strains is to give two different antibiotics so that each can inhibit mutants resistant to the other. This tactic is less useful against infectious spread of drug-resistance factors in a bacterial population, since these factors often reduce the sensitivity to several antibiotics simultaneously.

The mechanisms of development of resistance are as varied as the mechanisms of action of antibiotics themselves. In general, however, these mechanisms include the development of impermeability to the drug, change in the binding of drugs to enzymes that they inhibit and induction of enzymes which can lead to modification or destruction of the drugs themselves.

1.4 Pharmacological Considerations

The mechanisms of action of individual classes of antimicrobials are considered in separate chapters, and in each chapter there is also a discussion of the pharmacokinetic characteristics of the compounds. In order that the drug responses are more readily understood, certain general principles of drug action and use are described here.

Pharmacology is the study of drugs. In relation to antimicrobial drugs, the *pharmacodynamics* of the compounds may be considered as the mechanism of action, in other words, what the drug does to the microorganism, or to the host when adverse effects are being considered. The *pharmacokinetics* of antimicrobial drugs comprise those properties of the drug which determine the way in which the body handles the drug, or the actions of the body on the drug.

1.4.1 Routes of Drug Administration and Drug Absorption

Giving a drug by mouth is the most convenient way of drug administration. For oral administration to be effective, the site of action of the drug must either be in the gastro-intestinal tract (in which case absorption is not necessary or desirable), or the drug must be lipid-soluble so that absorption from the gastro-intestinal tract can occur. *Lipid-solubility* is a very important property if the drug is to be taken orally (for systemic action). Substances that are not lipid-soluble may be only partly or irregularly absorbed from the gut, and such substances are usually given parenterally (i.e. by-passing the gut) by injection.

Substances which are not stable to the action of stomach acid (for example benzyl penicillin) are also poorly absorbed if given orally. The presence of food can delay or stop the absorption of drugs, and thus it is usually preferable to give them before, rather than shortly after, food. Drugs which decrease gut motility, such as morphine and opiates, barbiturates and atropine-like anticholinergic agents, can also delay and decrease drug absorption from the gastro-intestinal tract. Most drug absorption occurs by passive diffusion in the small intestine which has a very large area for absorption.

Lipid solubility of a drug can be affected by its acidic or basic characteristics, and the pH of the environment in which the drug is found. The pK_a value for a particular drug is the pH at which 50 per cent of the drug molecules are in an ionised form, and 50 per cent are in a non-ionised form. The non-ionised form of the drug is more lipid-soluble than the ionised form; thus for good absorption to occur it is best for the drug to be in a medium in which it is not ionised. (The converse is true when drug excretion is being considered: the ionised form of a drug is more water-soluble, and thus it is of advantage for easy excretion to maintain the drug in an ionised form.) Compounds derived from acids or with acidic groups on the molecule usually have pK_a values on the acid side of pH7, and substances derived from bases usually have pK_a values on the basic side of pH7. If the pK_a of a substance is known, and the pH of the medium in which it is found is also known, then it is possible to predict whether the substance will be mainly in an ionised or unionised form, and hence whether the substance is more likely to be lipid-soluble or water-soluble. Thus substances derived from acids will be less· ionised at acid pH and highly ionised at basic pH, while for those derived from bases the reverse is true.

Parenteral injections are used to administer drugs which cannot be given by another route, and more specifically intravenous injection is used when a rapid onset of action is required, because the absorption step of drug intake into the body is circumvented. Following intra-muscular or subcutaneous injection, absorption of a drug from the injection site depends on its lipid-solubility and the blood supply to the injection site. Intrathecal injections are occasionally made in life-threatening situations, or when a rapid onset of action in the central nervous system is required.

Drugs may be applied topically to the skin or to the eye when a purely local action is desired, but whereas high local concentrations can be achieved in this way, local adverse effects and systemic absorption can occur and may thus limit the usefulness of this mode of drug administration.

When a drug which acts on the respiratory system is given by inhalation it must either be aerosolised or in the form of fine particles, so that it penetrates into the small bronchi. If the bronchi are blocked, inhalation is inappropriate as the drug may not reach its site of action. Careful attention must be given to the dose administered by this route, as systemic absorption may occur and cause toxic responses.

The formulation of a drug may affect the ease with which it can be absorbed; the absorption of a drug can be increased if it is 'micro-fined', or slowed by combining the drug with a molecule which must be removed in the body to release the active drug.

1.4.2 Drug Distribution

The lipid-solubility and degree of ionisation of a drug can affect the distribution of the drug in the body, for several biological membranes may have to be crossed before a drug reaches its site of action. One of these is the 'blood-brain barrier' which behaves as a lipid interface which lipid-soluble drugs cross easily, while lipid-insoluble drugs do not. The permeability of the blood-brain barrier is increased when the meninges are inflamed, which may be partly due to increased blood flow.

Effective concentrations of a lipid-insoluble drug may occur in tissues which the compound normally does not reach if high plasma concentrations are achieved by intravenous injection, but toxic effects may also become more pronounced. A good blood supply is essential if a drug is to reach its target site in sufficient concentration to be effective. To clear an infection in a tissue with a relatively poor blood supply such as bone, it may be justifiable to use higher than the normal doses of drug.

Drugs are carried in the blood either in physical solution in the plasma, or in combination with constituents of blood such as the blood corpuscles and plasma proteins. Any compound which is in free solution in plasma is available for transfer to the site of action, but drugs which are combined with other blood constituents may not be totally available. In such cases, there is usually an equilibrium between the drug in free solution and that combined with blood constituents. The most common blood constituents with which drugs are combined are the plasma proteins, and this phenomenon is called 'plasma binding' or 'protein binding'. Plasma-protein binding can be regarded as providing a reserve of drug, which replenishes the drug in physical solution and may thus affect the time over which effective concentrations are maintained. When two drugs are given simultaneously, and both compete for the same plasma-protein binding site, the one with the higher affinity will be preferentially bound to plasma protein, and more of the other drug will be available in a free form. In such circumstances, abnormally high or toxic concentrations of the unbound drug may occur. Examples of such an interaction include nalidixic acid displacing warfarin from plasma proteins resulting in prolongation of clotting time and haemorrhage, and sulphonamides displacing bilirubin from plasma proteins in newborn infants causing kernicterus (see Sections 2.7 and 4.4).

The concentration of an antimicrobial in plasma is a good indication of whether the dose is sufficient, and the effective ranges of plasma concentration have been established for many antimicrobial drugs.

1.4.3 Metabolism

Metabolic conversion may increase or decrease the activity of a drug. The major types of reaction involved are oxidation, reduction, hydroxylation and conjugation. Most of these processes are effected by the liver microsomal enzymes. (Microsomes are artefacts of homogenisation and centrifugation, and probably represent pinched-off vesicles of endoplasmic reticulum containing cytoplasm.) The metabolism of drugs usually makes them more readily excreted because the products of metabolism tend to be more water-soluble and more readily ionised than the parent compound. Examples of antimicrobials whose activity is changed by metabolism include the hydrolysis of erythromycin estolate to give free active erythromycin, and glucuronidation of chloramphenicol to an inactive product. Alterations in liver function can have major consequences in terms of drug metabolism; thus metabolism is slowed in liver failure, or accelerated if the microsomal enzymes have been induced with other drugs such as barbiturates, chloral hydrate, meprobamate, phenylbutazone or ethanol.

If liver function is poor, the concentration of plasma proteins is often decreased, and this may result in a higher concentration of free drug in the plasma.

1.4.4 Excretion

Excretion of drugs from the body usually occurs in the urine or by secretion into the bile. Other routes of excretion such as the lungs and the skin are not significant in the excretion of antimicrobials.

Drugs that are metabolised to a more water-soluble form and those which are more highly ionised (at physiological pH) than the parent compound, are more readily excreted via the kidney. The major mechanisms of excretion through the kidney are filtration at the glomerulus and secretion into the lumen of the kidney tubules by active transport. Measurement of creatinine clearance capacity gives a good indication of the functional state of the glomerular filtration process. Only compounds in free solution in the plasma are available for excretion via the kidneys.

For ionisable compounds, the pK_a value of a drug (or metabolite) and the pH of the glomerular filtrate can influence the rate of excretion of the drug (see Section 1.4.1).

Penicillin is excreted by active secretion into the lumen of the kidney tubules. Such active secretion of a drug into the urine can be disadvantageous, as it may be necessary to administer doses frequently in order to maintain therapeutic concentrations. The rate of penicillin excretion can be reduced by probenecid, a drug which slows the active transport of penicillin across the renal tubules.

Drugs which are excreted in the bile frequently have a molecular weight in excess of 300, and thus many substances which become glucuronidated are excreted by this route. Active secretion occurs in the liver, the bile is collected by ducts and then passes via the gall bladder into the small intestine. Unless reabsorption of the drug occurs during its passage through the small and large intestines, it is excreted in the faeces. If some reabsorption of the drug occurs at these sites, the drug may become available to exert its biological effects again, before it is once more secreted into the bile by the liver. This process is called *enterohepatic circulation*, and a drug molecule could undergo several such secretion/reabsorption cycles before it is finally lost from the body.

1.4.5 The Effects of Age, Liver Function and Kidney Function on Drug Response

The quantity of drug administered to an individual is usually based on the dose which has empirically been found to be effective in 'the normal 70 kg man'. Such a standard man seems to be a good guide in determining drug dosage, but such doses may be inappropriate in many cases for several reasons. Extremes of age and extremes of body weight are obvious examples, as are the effects of changes in liver and kidney function.

Age. In the very young, many biological functions are poorly developed. Acid secretion in the stomach, microsomal enzyme activity in the liver, glomerular filtration and tubular secretion in the kidney are all poorly developed in the premature baby and in the newly born. In addition, the blood-brain barrier does not appear to be fully functional very early in life. In the very young, therefore, the dose of drug must be reduced. In older children and young adults, drug dosage is usually determined on the basis of the body weight as a fraction of the 70 kg man. The elderly generally have a reduced ability to secrete stomach acid, as well as some decrease of liver and kidney function. The result may be an increased ability to absorb drugs which are acid-labile, and a decreased ability to metabolise and excrete them, all possibly contributing to higher than normal plasma concentrations which are encountered in the elderly if drug dosage is not adjusted.

Liver Function. The inability to metabolise a drug, or to secrete it into the bile, may cause it to accumulate in the body. As the toxicity of some drugs is related to their concentration in the plasma, toxic responses may be promoted by decreased liver function. Altered drug response during liver failure may also be caused by the decrease in plasma proteins which occurs during liver failure, when abnormally high concentrations of the free drug may occur in the plasma. In such circumstances it may be necessary to reduce the amount of the drug given and increase the frequency of administration. Monitoring of the plasma concentration of drugs may be essential under such circumstances.

Kidney Function. The urinary excretion of drugs or their metabolites can be very significantly delayed during kidney failure, and if the frequency of administration is not reduced then toxic effects may rapidly develop. The quantity administered at any one time during kidney failure should, however, not be reduced as this could result in inadequate plasma concentrations.

The duration of action of a drug in the body depends on the physicochemical characteristics of the drug, the formulation, the route of administration, and not least the way in which the drug is handled in the body. The half-life (or plasma half-time; plasma $t_{1/2}$) is a measure of the time it takes for the plasma concentration of a drug to be reduced by 50 per cent. The half-time must not be assumed to be the same as the duration of action of the drug. Drugs which become irreversibly bound to tissues can produce effects long after the time when they are undetectable in plasma.

Measurement of plasma concentration is the most generally applicable method available for estimating the amount of drug in the body. Comparison of plasma level of an antimicrobial, and its MIC (median inhibitory concentration) for a pathogenic microorganism provides a good indication of whether the antimicrobial will effectively combat the infection.

1.4.6 Side Effects and Toxicity

Antimicrobial drugs are given to help eliminate an infecting microorganism. The desired action of such a drug is to kill or incapacitate the microorganism; any effect on the host is regarded as a side effect, and can contribute to an undesirable toxicity of the compound towards the host.

Host toxicity can take many forms which are described in greater detail for individual antimicrobials in subsequent chapters. A generalised distinction can, however, be drawn between hypersensitivity reactions and other toxic effects. *Hypersensitivity reactions* occur in persons who have (possibly unknown to themselves) had previous contact with a drug. The hypersensitivity reaction (allergy) involves an antigen-antibody reaction. It is generally believed that certain drugs (or unidentified constituents of the drug preparation) or their products, act as antigens. Antibodies to them are produced in the body, so that on further exposure an antigen-antibody reaction occurs. The response which occurs can vary from mild rashes to a fatal anaphylactoid reaction. A characteristic of hypersensitivity reactions is that they are independent of the dose of drug administered. It is prudent to err on the side of caution when hypersensitivity is suspected and immediately withdraw the drug, and to ensure that both patient and those who have care of him, know that the drug or any close congeners must not be used in future.

Other forms of toxicity, i.e. those which are not regarded as hypersensitivity reactions, are frequently dose-dependent, and thus it is possible to predict when toxic effects are likely. It should therefore be a relatively simple matter to ensure that toxic levels of the drug are avoided. However, certain forms of drug toxicity are totally unpredictable and are regarded as idiosyncratic responses; fortunately such reactions are rare, but nonetheless they can be very serious when they do occur. Toxic concentrations of drugs are particularly liable to occur when the ability of the body to eliminate drugs is impaired, as in liver or kidney failure.

During oral administration of certain broad-spectrum antibiotics (e.g. tetracycline) most or all of some bacterial species may be killed. When this occurs, other species in the gastro-intestinal tract which are normally not pathogenic may multiply in an uncontrolled manner and result in *super-infection*. This usually involves diarrhoea, and can be life-threatening if not treated. In such circumstances it is necessary to identify the species responsible for the super-infection, and treat the patient with a drug to which this species is sensitive.

1.4.7 Drug Interactions

If two or more drugs are given together, and at some stage they share a common site of action, then the activity of one or both of the compounds may be changed. This action at a common site can lead to either increased or decreased response, and examples of drug interaction have already been given in the preceding sections. These are briefly summarised below, and some additional mechanisms are also discussed.

Interactions during absorption:

(1) Drugs altering gastro-intestinal motility and thus absorption of other drugs, e.g. atropine-like compounds, opiates, barbiturates.

(2) Drugs causing chelation of ions, e.g. tetracyclines, which chelate Ca^{2+} and Fe^{2+}.

(3) The presence of food delaying drug absorption.

(4) Alteration of pH and its effect on ionisation and lipid-solubility of drugs.

Interactions during distribution:

(1) Competition for plasma-protein binding sites, e.g. sulphona-mides displace tolbutamide and cause exaggerated hypogly-caemic responses.

Interactions with drug metabolism:

(1) Inhibition of drug metabolism during liver failure, and by monoamine oxidase inhibitors, leading to exaggerated responses.
(2) Induction of microsomal enzymes by barbiturates and pheny-toin leading to more rapid metabolism of drugs.

Interactions during excretion:

(1) The effects of altering urinary pH on the ionisation and hence excretion of drugs.
(2) Inhibition of active tubular secretion of penicillin by probene-cid.

Two terms which are commonly used during discussion of drug interactions are *synergism* and *antagonism*.

If a drug interaction leads to a response which is greater than that due to only one of the drugs acting by itself then the interaction can be regarded as being *synergistic*. In other words the drugs are inter-acting to give an overall increased response, whether this is a result of potentiation of effect or of addition of effects.

If a drug interaction leads to a response which is less than that due to one of the compounds acting by itself, then the interaction is *antago-nistic*. This can be as a result of competitive or non-competitive pro-cesses. An important example of synergism is the interaction of sulpha-methoxazole and trimethoprim in the combination known as cotrimox-azole (see 2.4 and 2.7). Sulphamethoxazole and trimethoprim act at successive steps in the intermediary metabolism of folic acid in bacteria. Given by themselves, high doses of each are needed in order to elimi-nate bacterial infection. When given together however, the dose of each drug can be greatly reduced to obtain the same chemotherapeutic effect. Such an interaction is illustrated in Figure 1.1 and is a good example of a synergistic drug interaction. Synergism can be of thera-peutic significance as the toxicity of the lower dosages of the two drugs used is much lower than that of an equieffective higher dose of only one of the compounds.

Figure 1.1: An Isobologram* of the Chemotherapeutic Effect of a Combination of Sulphadiazine and Trimethoprim in Mice Infected with *Proteus vulgaris*

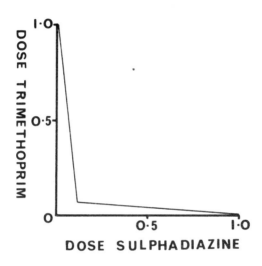

*An isobologram joins points at which a combination of drugs is equieffective to a full dose of one of the drugs alone. Thus, in this case, a combination of 12% of the full dose of sulphadiazine and 8% of the full dose of trimethoprim was as effective in curing infected mice as full doses of either drug given alone. If the effects of the drugs had been purely additive, then a line joining the 1.0 points of both axes would be expected.

Source: Reproduced with permission from Hitchings, G.H. and Burchall, J. (1965) *Advances in Enzymology 27*, 417-28.

2 SULPHONAMIDES AND DRUGS THAT INTERACT WITH FOLIC ACID

2.1 The Discovery of Sulphonamides

The studies of Ehrlich on the selective binding of dyes to microorganisms led to the discovery of sulphonamide drugs. **Prontosil rubrum**, a derivative of the dye 'chrysoidin' was shown in 1932 by Domagk to prevent the death of mice injected with a haemolytic streptococcus. Prontosil rubrum was found to be effective only *in vivo* and under the name of Streptozon it was introduced as a chemotherapeutic agent in 1935. The next year, Bovet showed that a reduction product of prontosil rubrum, namely **sulphanilamide**, was effective as a bacterial inhibitor *in vitro* as well as *in vivo*. The reduction of prontosil to sulphanilamide occurs *in vivo* in the liver. It was therefore concluded

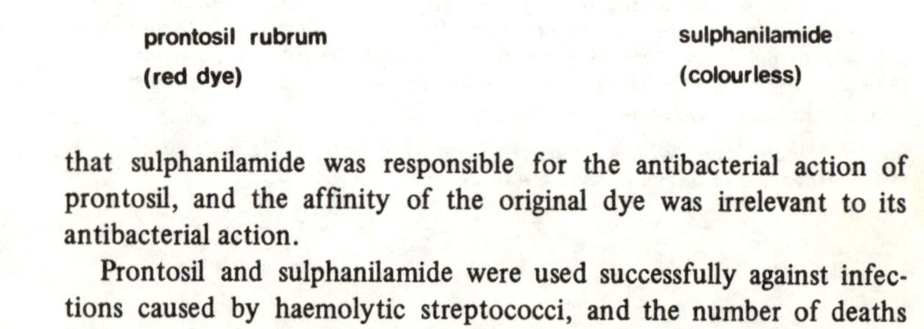

prontosil rubrum
(red dye)

sulphanilamide
(colourless)

that sulphanilamide was responsible for the antibacterial action of prontosil, and the affinity of the original dye was irrelevant to its antibacterial action.

Prontosil and sulphanilamide were used successfully against infections caused by haemolytic streptococci, and the number of deaths from puerperal sepsis (about 1,000 per annum in 1935 in England and Wales) was dramatically reduced.

The antibacterial action of sulphanilamide was found to be competitively inhibited by addition of extracts of bacteria or yeast to cultures of sulphanilamide-sensitive bacteria. It was suggested that the compound which competitively inhibited the action of sulphanilamide was a carboxylic acid of similar structure to sulphanilamide. The first synthetic compound tested as an 'anti-sulphanilamide' was **para-aminobenzoic acid (pABA)**, which was found to be highly effective. Woods (1940) suggested that pABA must be an essential metabolite involved in some unknown reaction, and that sulphanilamide was a competitive inhibitor of this process.

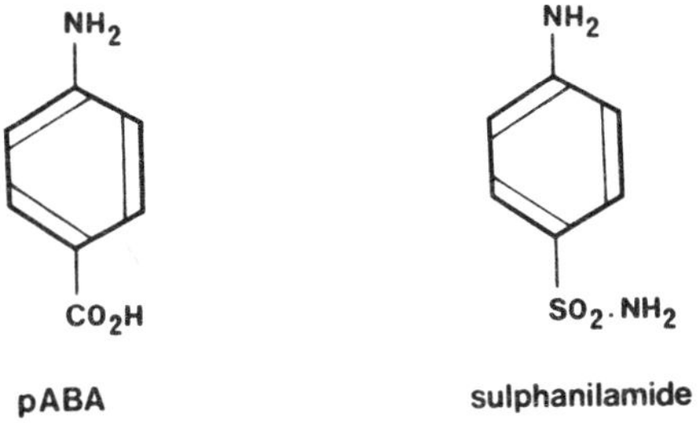

pABA   sulphanilamide

2.2 The Mechanism of Action of Sulphonamides

Sulphonamides inhibit the incorporation of pABA into a precursor of

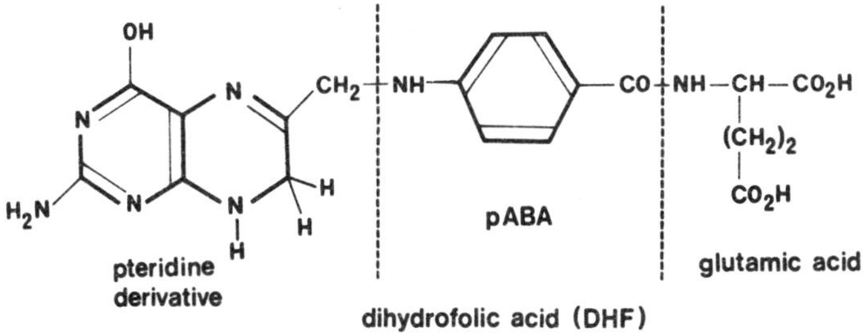

dihydrofolic acid (DHF)

dihydrofolic acid (DHF) which is then reduced by the enzyme *dihydrofolate reductase* to *tetrahydrofolic acid*. This derivative of folic acid

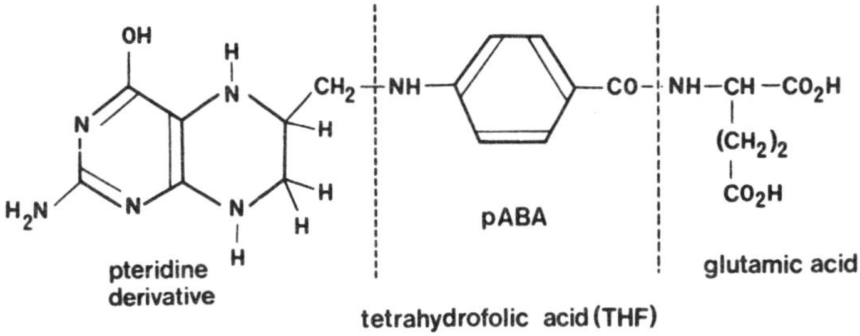

tetrahydrofolic acid (THF)

25

is an important coenzyme involved in the transfer of small residues containing a single carbon atom (e.g. methyl, formyl) in intermediary metabolism. The synthesis of the amino acid methionine and of nucleic acid bases involves the use of THF derivatives (see Figure 2.1)

Figure 2.1: The Biosynthetic Uses of Folic Acid*

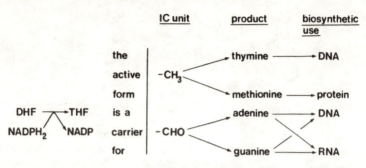

*Dihydrofolate (DHF) is reduced to tetrahydrofolate (THF), which is the active carrier coenzyme for many one-carbon fragments such as methyl and formyl groups. These fragments are incorporated into proteins via methionine, and into DNA and RNA via the nucleotide bases.

Dihydrofolic acid is synthesised in two stages, of which the first is the combination of the pteridine derivative with pABA by *dihydropteroic acid synthetase*, the reaction that is inhibited by sulphonamides. This is followed by condensation of glutamic acid with dihydropteroic acid. Because pABA and sulphonamides compete for the same enzyme, the inhibitory effects of sulphonamides can be reversed if exogenous pABA is available in high enough concentrations. Unlike man, bacteria (with very few exceptions) do not utilise exogenous folates, which they cannot absorb (see Figure 2.2). Bacteria therefore depend on the synthesis of dihydrofolate within their cells, and if this process is inhibited the formation of macromolecules is impared. Other drugs, e.g. trimethoprim (see Section 2.4), which inhibit the conversion of dihydrofolate to tetrahydrofolate, also limit the supply of these biosynthetic components. Therefore, the time lag between administering sulphonamides and the cessation of bacterial growth corresponds to the time taken to use up the stocks of these biosynthetic components, and of folic acid already present in the cell. Growth proceeds for about four cell generations after sulphonamide treatment, by which time the original stock of folic acid is diluted amongst the progeny of the

Figure 2.2: The Effects of Sulphonamides and Trimethoprim on Folic Acid Metabolism in Bacteria

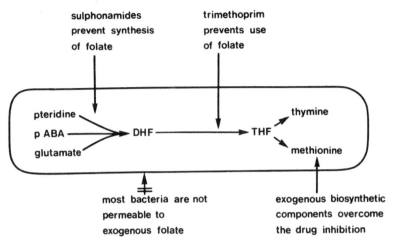

treated bacteria. Because the sulphonamides inhibit the synthesis of nucleotides and amino acids, the drugs are not effective if these compounds are available to the bacteria from exogenous sources (Figure 2.2), e.g. when there is extensive tissue damage in cases with burns and purulent exudates. Most tissue fluids, including blood, have relatively low concentrations of amino acids and do not contain nucleotide bases in a form that can be absorbed by bacteria.

The sulphonamides are synthetic derivatives of sulphanilamide formed by substitution at N4 (e.g. **sulphamethoxazole** and **sulphafurazole**) and are not therefore antibiotics. The amino group (N1) is usually

sulphamethoxazole

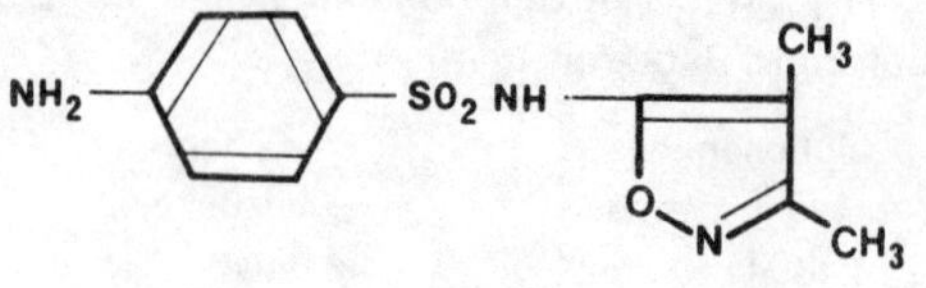

sulphafurazole

unsubstituted except in the case of succinyl and phthalyl sulphona-mides (e.g. succinyl sulphathiazole) which have a long duration of action and are not perceptibly absorbed from the gut. Active sulphona-mide is slowly released due to hydrolytic removal of the succinyl or phthalyl residue.

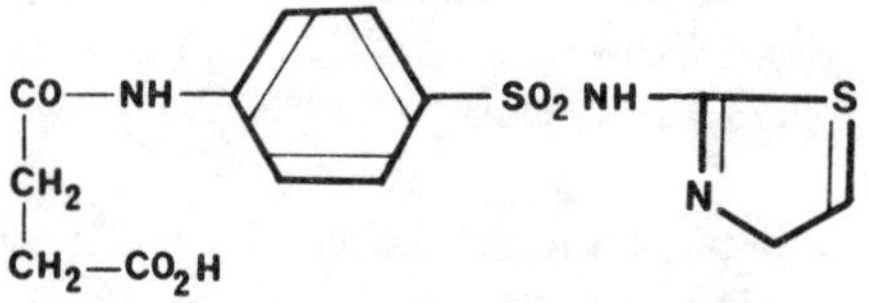

succinyl sulphathiazole

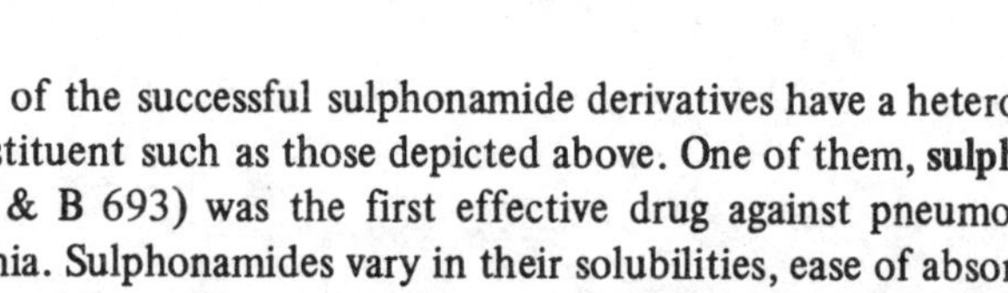

phthalyl sulphathiazole

Many of the successful sulphonamide derivatives have a heterocyclic ring substituent such as those depicted above. One of them, **sulphapyri-dine (M & B 693)** was the first effective drug against pneumococcal pneumonia. Sulphonamides vary in their solubilities, ease of absorption, metabolic fate and excretion (see Section 2.6). The sulphonamides were initially used for streptococcal infections and pneumococcal pneumonia, for which they have been largely replaced by natural antibiotics. As a high proportion of the dose of certain sulphonamides (e.g. sulphafurazole) is excreted unchanged in the urine, they still have a role in the treatment of urinary tract infections.

In *Escherichia coli*, sulphonamides may also act as substrates for the enzyme dihydropteroic acid synthetase (see p. 26) and produce folic acid analogues. These compounds may exert their own inhibitory

dihydropteroic acid analogue

effects, but their prevalence and significance is not clear. However, as with many other antimicrobial agents, it is too simplistic to assume that there is only one mode of action, and that this is the same in all bacteria.

2.3 Other Folic Acid Antimetabolites

The search for useful chemotherapeutic agents has involved the synthesis not only of a large range of sulphonamides but has also extended to other competitors of p-aminobenzoic acid. The growth of the tubercle bacillus *Mycobacterium tuberculosis* is inhibited by **p-aminosalicylic acid** (pAS), and most strains are sensitive *in vitro* to concentrations as low as 1 μg/ml. The bacteristatic effect is specific to this microorganism, and pAS is thought to inhibit folic acid synthesis as an analogue of p-aminobenzoic acid. The specificity implies an ability of the target enzyme to bind pAS that is not found in other microorganisms. Chemotherapy of tuberculosis required 8-12 g per day of pAS, or less if it was combined with other drugs such as isoniazid (isonicotinic acid hydrazide). The advent of effective antibiotics (streptomycin, rifampicin) has now reduced the use of pAS.

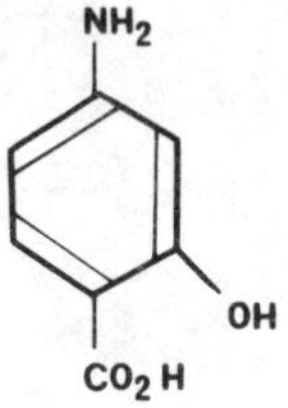

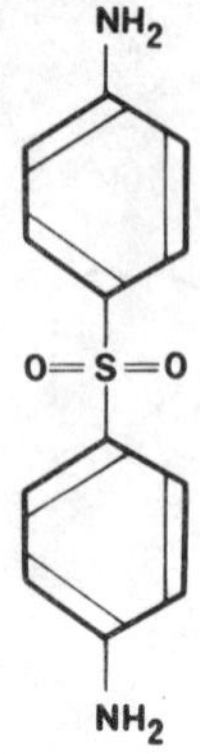

pAS

(M. tuberculosis)

Dapsone

(M. leprae)

By contrast with pAS the sulphone drug dapsone (diaminodiphenyl-sulphone) remains the best available drug for treating leprosy. Although more effective against streptococcal infections than sulphanilamide, dapsone was also more toxic and was considered unlikely to be useful clincially. However, dapsone was able to suppress experimental tuberculosis infections and most strains of *M. tuberculosis* are inhibited by 10 µg/ml *in vitro*. Clinical trials of the use of dapsone in leprosy were made in the 1940s, but investigation of the mode of action is hampered because *Mycobacterium leprae* cannot be grown adequately *in vitro*. The mode of action is presumed to be similar to that of the sulphonamides and pAS.

2.4 Antagonists of Dihydrofolate Reductase

The enzyme (dihydrofolate reductase) which reduces dihydrofolate to tetrahydrofolate, the active form of folic acid, is essential to all cells, whether they make folic acid (sulphonamide-sensitive bacteria) or require pre-synthesised folic acid (vitamin-requiring higher organisms). Analogues that closely resemble folic acid, e.g. **methotrexate**, are taken up by mammalian cells and inhibit the mammalian dihydrofolate reductase. Such folate analogues are not absorbed by bacteria and are therefore no use in the chemotherapy of bacterial infections,

30

although they have been applied to the treatment of leukemia.

methotrexate (antileukemia)
a close analogue of folic acid

Dihydrofolate reductases from different organisms do not have identical properties and surveys of a wide range of compounds have resulted in the identification of drugs that inhibit the enzyme in protozoa and bacteria but not in higher organisms. **Trimethoprim** is a pteridine analogue which binds to the bacterial dihydrofolic acid reductase

trimethoprim (antibacterial)
selectively inhibits bacterial
dihydrofolate reductase

preventing the conversion of DHF into the useful form, THF. Of the order of 50,000 times more trimethoprim is needed to inhibit mammalian enzymes than will block the bacterial enzyme, so that therapeutic concentrations achieved in the body are not hazardous to the host cells.

A combination of sulphamethoxazole and trimethoprim known as **cotrimoxazole** is better than either drug alone, presumably because they affect the same pathway at different sites (Figure 2.1). The synergistic effects of two drugs given in combination has been illustrated by the isobologram in Figure 1.1.

The trimethoprim serves to prevent the selection of sulphonamide-resistant mutants. Because of the rapid onset of activity of trimethoprim, the combination is more immediately effective than sulphonamides alone. The bactericidal spectrum of cotrimoxazole is also wider

31

than that of either of its constituents because the combination is able to stop growth of strains against which each is only marginally active.

2.5 Sensitive Bacteria and Resistance to Sulphonamides

Bacteria that are sensitive to the sulphonamides include certain strains of streptococci including pneumococci, some actinomycetes, *Corynebacterium diptheriae* and *Haemophilus influenzae*. Sulphonamides are still used to a limited extent for the treatment of bacterial infections, usually in combination with another drug. Cotrimoxazole (see above) is effective against *C. diptheriae*, *Steptococcus pneumoniae* and *Neisseria meningitidis* as well as many enteric microorganisms and is often used for urinary tract infections. Methicillin-resistant strains of *Staphylococcus aureus* are often sensitive to the combination of sulphamethoxazole and trimethoprim but not to either of them separately.

Examination of the role of sulphonamides in the competitive inhibition of the synthesis of folic acid (Figure 2.2) reveals that bacterial resistance to sulphonamides could be due to acquired impermeability of the bacterial membrane to the drugs, or to increased production of the substrate (pABA) to overcome the inhibition, or to a change in the properties of dihydropteroic acid synthetase so that the enzyme binds the inhibitor less readily. Both of these latter types of resistance have been demonstrated. The increased intracellular concentration of pABA in sulphonamide-resistant bacteria is presumably due to a loss of the normal end-product inhibition of pABA synthesis. An example of the changed properties of dihydropteroic acid synthetase has been demonstrated in a resistant mutant in which the enzyme has greater affinity for the substrate (lower K_m), and a reduced affinity for sulphonamides (higher K_i) than the normal sulphonamide-sensitive enzyme (see Table 2.1).

Table 2.1: Affinity Constants of Dihydropteroic Acid Synthetase of a Pneumococcus and its Sulphonamide-resistant Mutant

		Affinity constant (mM)	
Substrate		Wild pneumococcus	Resistant mutant
pABA	(K_m)	0.1	0.06
sulphonamide	(K_i)	0.5	65

Source: Wolf, B. and Hotchkiss, R.D. (1963) *Biochemistry 2*, 145.

2.6 Pharmacology

2.6.1 Routes of Administration

Sulphonamides are available for oral, parenteral and topical adminis-tration. Following oral dosage, sulphonamides which are absorbed from the gut have a duration of action which is determined by the speed of metabolism and excretion. **Sulphafurazole** is very water-soluble and is readily excreted in the urine, in which high concéntra-tions may be achieved. Sulphafurazole is therefore used for treatment of urinary tract infections. **Sulphadiazine** and **sulphadimidine** can be given orally or parenterally. Both they and sulphamethoxazole are well absorbed from the gastro-intestinal tract. Sulphamethoxazole is usually employed with trimethoprim in the combination known as cotrimoxazole (see above). **Phthalyl sulphathiazole** and **succinyl sulpha-thiazole** (see p. 28) are insignificantly absorbed from the gut, and can be used to 'sterilise' the gut prior to surgery.

Many sulphonamides are absorbed when applied to the skin, but they are not generally administered in this way as they are liable to cause skin sensitisation and rashes. **Sodium sulphacetamide** is not an irritant when applied in aqueous solution to the eye, and penetrates well into the occular fluid. **Silver sulphadiazine** is sometimes used topically to treat burns as it appears to be bactericidal to *Pseudomonas aeruginosa*.

2.6.2 Distribution and Duration of Action

Following absorption, sulphonamides are transported either in physical solution in the plasma, or bound to plasma proteins. At therapeutically-effective doses, sufficient concentrations of drug are free in the plasma for effective concentrations to build up in target tissues. Sulphonamides can enter most body compartments including the cerebrospinal fluid. Provided adequate blood supply is available to a tissue, the concentration of sulphonamide in the tissue compartment may be 50-80 per cent of the unbound concentration in the plasma. The permeability of the meninges to sulphonamides is increased when they are inflamed.

2.6.3 Metabolism

Sulphonamides are acetylated in the liver to chemotherapeutically less active, or inactive, acetylated derivatives. The acetylation occurs at amino groups with acetyl coenzyme A as the donor. The toxicity of sulphonamides is not lost, however, following metabolism, and indeed it may be increased. If the liver is damaged, acetylation may be slowed, and the half-life of a compound in the plasma increased. In such cases dose frequency should be decreased.

The ability to acetylate sulphonamides is not the same in all subjects, but whether a person is a fast or a slow acetylator is not significant in sulphonamide chemotherapy. However, the ability to acetylate **isoniazid** is important in determining optimum dosage of isoniazid for the treatment of tuberculosis. The acetylation of sulphadimidine is easily determined and is used as an indicator of ability to acetylate isoniazid, since this is not easily measured in man.

2.6.4 Excretion

Sulphonamides, together with their derivatives, are excreted in the urine. The rates of excretion of individual sulphonamides depend on their physicochemical properties and on kidney function. Sulphonamides and their metabolites can accumulate in kidney failure and the dose must be adjusted to control the incidence of toxicity. It is desirable to maintain high fluid intake during sulphonamide chemotherapy in order to reduce the incidence of kidney toxicity. Even the highly water-soluble sulphonamides which are currently favoured for systemic use, may crystallise in the urine if the flow is small. Acetylation of sulphonamides increases their acidity. The parent compounds are usually neutral and therefore the acetylated derivatives (which are more acidic) are more highly ionised at physiological pH. The degree of ionisation and therefore the rate of excretion can be increased further by making the glomerular filtrate, and therefore the urine, more alkaline by giving sodium bicarbonate.

Mixtures of three sulphonamides are occasionally used in order to decrease the danger of precipitation of crystals in the urine. The solubility in urine of each compound is independent of the concentration of other compounds, whereas the chemotherapeutic effect of such mixtures is the sum of those of the three constituents.

2.7 Toxicity

Adverse reactions to sulphonamides occur in about five per cent of the population, but minor side effects should not be ignored as they may indicate that a more serious reaction may follow if the drug is not discontinued. An adverse response to one sulphonamide usually means that there will be further adverse reactions to an alternative sulphonamide, or if sulphonamides are used on a subsequent occasion. A history of sulphonamide-sensitivity therefore contraindicates their use. Because rapidly-excreted sulphonamides have side effects of shorter duration, it seems prudent to choose such a compound when the sensitivity of an individual is not known.

Fever induced by sulphonamides can occur, usually after about one week of treatment, and thus the course of treatment must be closely supervised. Skin reactions in the form of rashes, itching and irritation of mucous membranes can occur with any sulphonamide in a sensitive individual, and incidence of fever and skin reactions increases with higher doses.

The risk of precipitation of crystals of sulphonamides or their metabolites in the urine (crystaluria) can be reduced by using compounds which themselves are highly water-soluble and which do not give rise to water-insoluble metabolites, by ensuring that adequate water intake is maintained, and by not using sulphonamide drugs when kidney function is impaired. Hypersensitivity reactions to sulphonamides can manifest themselves as impairment of kidney function; thus renal function should be monitored during sulphonamide chemotherapy. Making the urine alkaline reduces the incidence of renal damage.

Sulphonamides may on rare occasions adversely affect blood-forming

historicaly
historically

tissues. Such effects tend to be unrelated to drug levels and may occur after the course of drug treatment is completed. Sulphadiazine is more frequently responsible for acute haemolytic anaemia and agranulocytosis than are other sulphonamides. Other blood disorders such as aplastic anaemia, thrombocytopoenia and eosinophilia may be caused by sulphonamides.

Cotrimoxazole is a mixture of one part trimethoprim and five parts of sulphamethoxazole. The mixture is well absorbed from the gastrointestinal tract and the combination gives the optimum synergistic (see Section 1.4.7) plasma levels of the two compounds. Trimethoprim is largely excreted unmetabolised in the urine, and as it is a weak base, excretion is particularly rapid if the urine is acidic.

The side effects of trimethoprim are similar to those of sulphonamides, and as trimethoprim is not available except in combination with sulphamethoxazole, only disturbances of folate metabolism can be attributed unambiguously to trimethoprim. These are manifested chiefly as blood disorders such as reduced erythropoiesis, leucopoenia and thrombocytopoenia. Normal blood-cell development can be reestablished by giving folinic acid (formyl-THF) to patients who show these side effects.

Interactions

The rate of excretion of sulphonamides by the kidney can be significantly increased if the urine is made more alkaline by giving sodium bicarbonate.

The percentage of the total drug bound to plasma protein varies amongst the sulphonamides. Some appear to have a high affinity for the plasma-protein binding sites to which tolbutamide and bilirubin are bound. In the case of the orally administered hypoglycaemic drug tolbutamide, simultaneous administration of a sulphonamide can lead to an earlier and more profound hypoglycaemia, which may also last for longer than when tolbutamide alone is used. In the case of bilirubin in the neonate, sulphonamides can cause jaundice by displacing the bilirubin from the plasma protein binding sites. The increase in free plasma bilirubin in the neonate is known as kernicterus, and can occur if the mother is given sulphonamides during pregnancy, or if sulphonamides are given to the infant before bilirubin excretion is completely developed.

3 INHIBITORS OF PEPTIDOGLYCAN SYNTHESIS

3.1 History and Introduction to Antibiotics Active at Cell Walls

In 1928 Fleming noted that staphylococcus colonies on plates infected with a mould of the genus *Penicillium* became transparent and that the staphylococcus cells lysed. The culture medium in which the penicillin mould was grown was bactericidal to various cocci, but was ineffective against Gram-negative bacteria such as *Escherichia coli*. The name 'penicillin' was originally given to crude filtrates of the mould culture, which were found not to be toxic to man or animals. Inability to isolate the active ingredient of penicillin mould extract held up progress, but in 1941 a group headed by Chain and Florey was successful. Early products were probably mixtures of a number of penicillins, but with the development of deep-culture growth techniques, and the isolation of high-yield mutants, it has been possible to produce a number of different pure penicillins.

Gardner (1940) noted that sub-lethal concentrations of penicillin caused lengthening and swelling of sensitive bacteria, and Duguid (1946) proposed that penicillin interfered with normal cell-wall formation so as to produce weak walls that might burst with loss of cell contents. In 1949 Park described the accumulation of nucleotides containing amino acids in staphylococci treated with penicillin, and these compounds ('Park nucleotides') were later identified as precursors of cell-wall *peptidoglycan* (also called mucopeptide). The structure and synthesis of this important component has been elucidated in parallel with the identification of the site of action of several antibiotics described in this chapter.

3.2 Mechanism of Action on the Peptidoglycan Target

3.2.1 The Structure of the Bacterial Cell Wall and General Effects

The cell membrane of bacteria is a delicate structure which surrounds a cytoplasm with relatively high concentrations of low molecular weight substrates, and which therefore has a high osmotic pressure. Like plant cells, bacteria have an exterior cell wall against which the

37

cell membrane is applied by turgor pressure. The major structural feature of the bacterial cell wall from the standpoint of mechanical strength is peptidoglycan (mucopeptide) which is found in both Gram-positive and Gram-negative bacteria.

Gram-positive bacteria stain purple with Grams stain. They often need many essential nutrients, which are supplied in complex growth media. They have high internal concentrations of metabolites causing internal osmotic pressures as high as 25 atmospheres. The cell walls are 200-500 Å thick and relatively simple, comprising a layer of peptidoglycan next to the cell membrane which contributes about 50 per cent of the total mass of the walls. Other components of the wall include proteins, polysaccharides and teichoic acids. Gram-negative bacteria fail to stain purple with Grams stains and are observed to be pink due to the counterstain. The Gram-negative bacteria are often biosynthetically versatile, and therefore need few or no essential nutrients. The cytoplasm has low concentrations of metabolites so that the internal osmotic pressure may be as low as five atmospheres. The cell wall is thin (100-150 Å), but complex, and more than one layer can be seen in the electron microscope. Peptidoglycan may constitute as little as 1-10 per cent of the cell walls, which contain variable amounts of lipoprotein and lipopolysaccharide (endotoxin). Despite the low proportion of peptidoglycan in the Gram-negative wall, it is this component that determines the shape of the cells and confers rigidity and strength to the walls, as treatment with lytic enzymes that hydrolyse peptidoglycan causes loss of shape and integrity.

Failure to synthesise adequate peptidoglycan results in the growing bacteria swelling and lysing (bursting) due to their high internal osmotic pressure. Antibiotics that block peptidoglycan synthesis have no effect on the mature peptidoglycan of non-growing bacteria, and in growing cells their effect is only evident at the sites where new wall material is being intercalated into the existing structure. In bacilli, aberrant morphological forms (for example 'rabbit ear' shapes) may be seen. In these, the ends of the rods comprising 'old' mucopeptide retain their shape (the ears), but the central parts, where new synthesis is occurring, balloon out. These antibiotics are therefore bactericidal towards growing cells. By contrast, the enzyme *lysozyme* (which was also discovered by Fleming and which serves as a protective agent in the body fluids and secretions of animals), hydrolyses the β-1:4 bonds between the carbohydrate molecules (see Section 3.2.2) of mature peptidoglycan, and lyses both growing and non-growing cells.

3.2.2 Peptidoglycan Structure

When *Staphylococcus aureus* was treated with sub-lethal concentrations of penicillin, compounds containing a nucleotide, an amino sugar and several amino acids accumulated. These substances were proposed to be precursors of peptidoglycan. It was further discovered that cell walls synthesised in such a system contained more alanine than normal, and that glycine residues had free amino groups which is not normally the case. This points to failure of cross-linking of peptidoglycan as a basis for the action of penicillin, and is consistent with the effects of penicillin on cell morphology described above.

Peptidoglycan is a lattice-like macromolecule (see Figure 3.1)

Figure 3.1: The Structure of Peptidoglycan (Mucopeptide) of *Staphylococcus aureus**

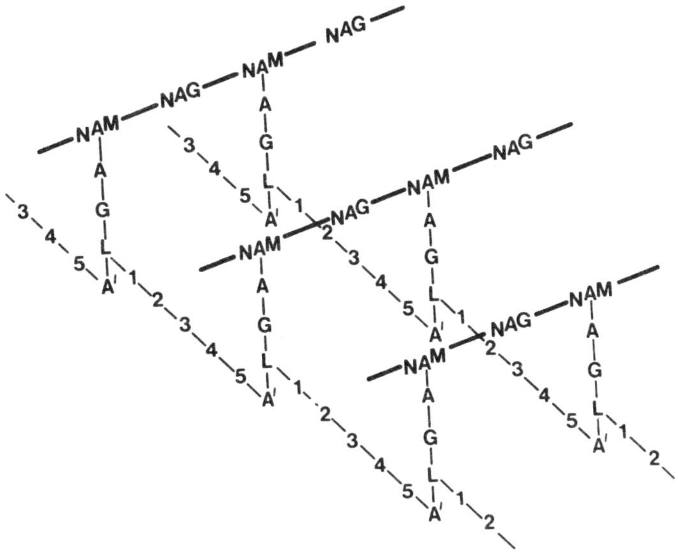

*Carbohydrate strands consisting of alternate *N*-acetylmuramic acid (NAM) and *N*-acetyl glucosamine (NAG) are linked via tetrapeptides (A-G-L-A') cross-linked by pentaglycine chains (1-2-3-4-5).

comprising polysaccharide (glycan) strands oriented parallel to one another and cross-linked by short peptides. The glycan chains vary little from bacterium to bacterium and comprise *N-acetylglucosamine*

(NAG) alternating with the 3-*0*-lactyl of *N*-acetylglucosamine (*N*-acetylmuramic acid, NAM), linked together by β-1:4 bonds. Peptide chains are attached to the glycan by peptide bonds between the amino

NAM **NAG**

group of the first amino acid and the carboxyl group of the lactic acid residue of NAM. This peptide is peculiar in its structure in that it contains alternating D and L amino acids. In mature peptidoglycan (mucopeptide) NAM is linked to a tetrapeptide (A-G-L-A', as in Figure 3.1) in which A is L-alanine, G is D-glutamic acid or D-glutamine, L is L-lysine or another dibasic amino acid and A' is D-alanine. Another peculiarity of the structure of this peptide is that the glutamic acid or glutamine residue is linked via its 5-carboxyl in a peptide bond rather than via its 1-carboxyl as in proteins. This unusual arrangement is distinguished by describing the residues as D-isoglutamic acid or D-isoglutamine. The tetrapeptides are cross-linked from the side chain

glutamic acid (or glutamine)   isoglutamic acid (or isoglutamine)
as in proteins   as in mucopeptide

amino group of the dibasic amino acid L to the terminal carboxyl of residue A′ of another tetrapeptide chain. The cross-link is between tetrapeptides attached to different glycan chains producing a network (Figure 3.1). In many cases the cross-link is direct, but in some instances bridges are composed of other amino acids. The peptidoglycan of *Staphylococcus aureus* has lysine linked to D-alanine by a pentapeptide (1-2-3-4-5 in Figure 3.1) comprising glycine residues (pentaglycine).

3.2.3 The Synthesis of Peptidoglycan

Peptidoglycan is made in several stages; the initial reactions occur in the cytosol of the cell (see Figure 3.2), further transformations are effected in the cell membrane (see Figure 3.3), and the final incorporation of peptidoglycan into the bacterial cell wall occurs at the

Figure 3.2: The Assembly of Peptidoglycan Precursor in the Bacterial Cytoplasm*

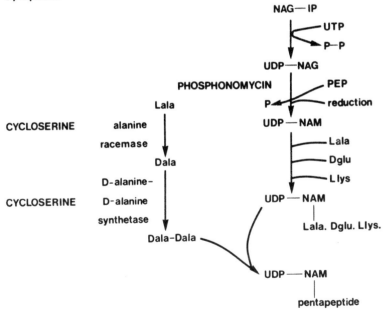

*The synthesis of NAM pentapeptide combined with UDP is inhibited by phosphonomycin and D-cycloserine.

Figure 3.3: The Synthesis and Transfer across the Membrane of the Complete Peptidoglycan Subunit*

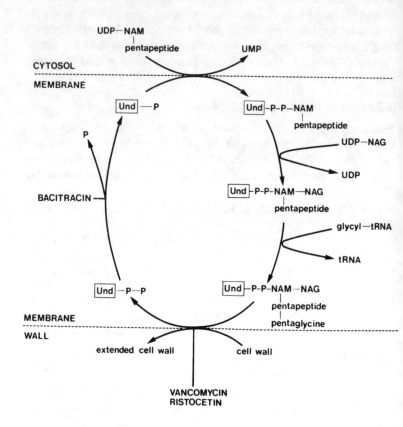

The UDP-activated cytoplasmic product is transferred to an undecaprenyl lipid carrier (Und-P) in the membrane. This intermediate receives *N*-acetyl glucosamine (NAG) from UDP-NAG, also of cytoplasmic origin. Pentaglycine is also added in the membrane and the entire subunit is transferred from undecaprenyl pyrophosphate onto a growing point in the peptidoglycan macromolecule. Vancomycin and ristocetin inhibit this transfer while bacitracin blocks the regeneration of undecaprenyl monophosphate from the pyrophosphate.

points of extension of the cell wall. The basic subunit of peptidoglycan made in the cytoplasm of the bacterial cell comprises NAM attached to the activating nucleotide uridine diphosphate (UDP) and bearing a pentapeptide (A-G-L-A'-A'), not a tetrapeptide (A-G-L-A') as in the mature peptidoglycan (Figure 3.2). *N*-acetyl glucosamine-1-phosphate is first converted into UDP-NAG. The lactic acid residue which converts UDP-NAG into UDP-NAM is incorporated as pyruvate from the glycolytic intermediate phosphoenolpyruvate (PEP), followed by reduction of the pyruvate to lactate. The carboxyl group of this lactic acid is the site of peptide attachment. The stepwise addition of three amino acids by peptide bonds with the consumption of ATP forms the tripeptide derivative UDP-NAM-tripeptide. The last two amino acids of the pentapeptide chain are added as the dipeptide D-alanine-D-alanine which is synthesised separately by a synthetase. D-alanine is produced from L-alanine by alanine racemase.

The next stage of peptidoglycan synthesis takes place in the cell membrane (Figure 3.3). The membrane carrier employed is a C_{55} lipid called undecaprenyl phosphate (Und-P) comprising eleven isoprene units. This accepts phospho-NAM-pentapeptide, releasing UMP into the cytoplasm. The lipid complex in the membrane accepts NAG from cytoplasmic UDP-NAG so that the growing peptidoglycan subunit now contains the NAM-NAG disaccharide. In the case of *Staphylococcus aureus* the pentaglycine side-chain is also combined with the lysine amino group (of A-G-L-A'-A') at this stage, the donor being a specific glycyl-tRNA distinct from that involved in protein biosynthesis (Figure 3.3).

Growing bacteria have active lytic enzymes that locally hydrolyse the mucopeptide of the cell wall to allow new components to be added at growing points. The peptidoglycan subunit is detached from the undecaprenyl pyrophosphate (Und-P-P) carrier and transferred to a growing point in the peptidoglycan by a bond to the NAM-NAG disaccharide. The released undecaprenyl lipid is in the pyrophosphate form, which must be hydrolysed by a specific pyrophosphatase in order to regenerate the monophosphate that accepts UDP-NAM-pentapeptide from the cytosol.

If peptidoglycan were synthesised only by the processes described so far, the glycan chains would be extended, but no lateral cross-links between them would be established. The process of cross-linkage by *transpeptidation* is illustrated in Figure 3.4. The side-chain amino group of the pentaglycine of one glycan chain reacts enzymically with the peptide bond between two D-alanine residues of a pentapeptide (described above as A-G-L-A'-A') from another glycan chain. The reaction involves the migration of the peptide bond and the transfer

Figure 3.4: Cross-linking of Peptidoglycan by Transpeptidase

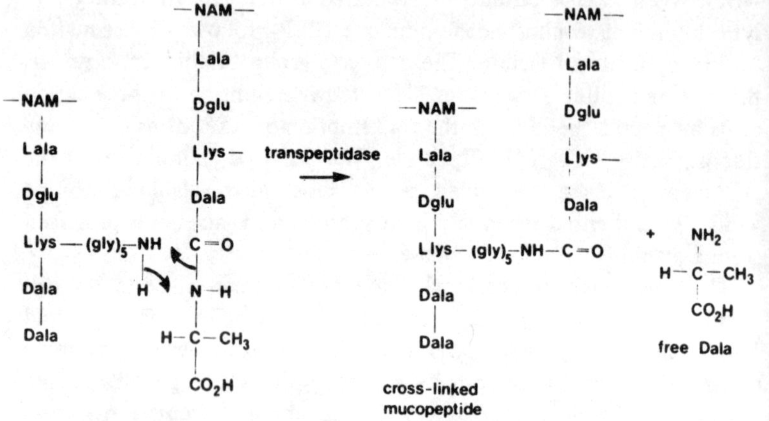

of a proton from the pentaglycine amino group; free D-alanine is released. The details of the reaction, and the extent to which cross-linkage occurs, vary from microorganism to microorganism.

3.2.4 Sites of Action of Individual Antibiotics

The processes in peptidoglycan biosynthesis that are affected by antibiotics have already been indicated in Figures 3.1, 3.2 and 3.3. **Phosphonomycin** is a recently-discovered antibiotic which has not been applied widely in clinical medicine although clinical trials have been carried out in Spain and Japan. Its structure somewhat resembles that of the glycolytic intermediate phosphoenolpyruvate (PEP), but the resemblance is not close, and phosphonomycin is not a general inhibitor of reactions involving PEP.

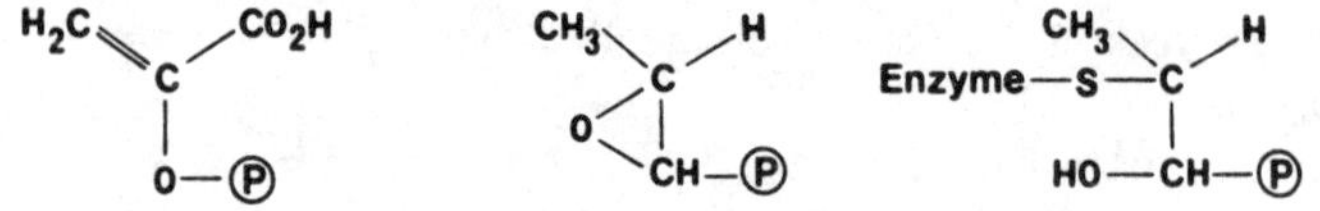

phosphoenolpyruvate   phosphonomycin   enzyme derivative via
                                                      cysteine

44

The transfer of pyruvate from PEP to NAG during peptidoglycan synthesis is specifically inhibited by phosphonomycin. However, the affinity of the antibiotic for the pyruvate transferase is not great, and high concentrations of the antibiotic have to be employed. The inhibition involves the reaction of phosphonomycin with a cysteine residue in the enzyme. The specificity of this reaction contributes to the low toxicity of phosphonomycin.

D-Cycloserine is used clinically as a reserve drug for the treatment of infections resistant to other antibiotics (e.g. tuberculosis). It is not a drug of first choice because of its deleterious effects on the nervous system. The molecule is an analogue of D-alanine and inhibits both the racemisation of L-alanine and the synthesis of the D-alanine dipeptide (Figure 3.2). In some bacteria, cycloserine causes the accumulation

D-alanine   D-cycloserine

of a 'Park nucleotide', UDP-NAM-tripeptide, which lacks the terminal D-alanine-D-alanine normally combined with it in synthesis. The inhibitory effects of cycloserine are reduced by provision of exogenous D-alanine in some strains.

Vancomycin causes the accumulation of 'Park nucleotides' in Gram-positive bacteria and inhibits the incorporation of amino acids into peptidoglycan. Various compounds that contain the D-alanine dipeptide, including UDP-NAM-pentapeptide, bind to vancomycin but this is unlikely to play a part in the mechanism of action as this compound is produced in the cytoplasm and vancomycin is not able to pass through the bacterial cell membrane. The binding of vancomycin to the D-alanine-D-alanine residue probably takes place at the membrane surface, so preventing the transfer of the subunit containing the dipeptide into the growing cell wall. **Ristocetin** is an antibiotic that has a similar mechanism of action. The toxicity of both of these drugs precludes their use except for severe infections with staphylococci which fail to respond to other antibiotics.

Bacitracin is a peptide containing linear and cyclic components. It is too toxic for systemic use but is sometimes applied topically as a local agent, e.g. to suppress the Gram-positive flora in surgery of the abdomen. Its mode of action is as an inhibitor of the membrane pyrophosphatase that releases undecaprenyl phosphate from undecaprenyl pyrophosphate. It therefore causes the accumulation of the lipid carrier as the pyrophosphate, so preventing the transfer of peptidoglycan precursors from UDP derivatives in the cytoplasm to undecaprenyl monophosphate in the membrane.

Penicillins are acid (R.CO-) derivatives of the penicillin nucleus (6-aminopenicillanic acid) which is derived from cysteine and valine

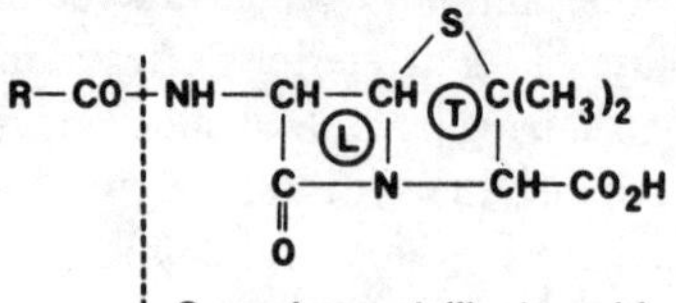

6 - aminopenicillanic acid

(L) = β - lactam ring

(T) = thiazolidine ring

condensed together to form a rigid double-ring system. The bottom edge of the penicillin molecule resembles the sequence of the terminal dipeptide of uncross-linked mucopeptide, D-alanine-D-alanine, which is

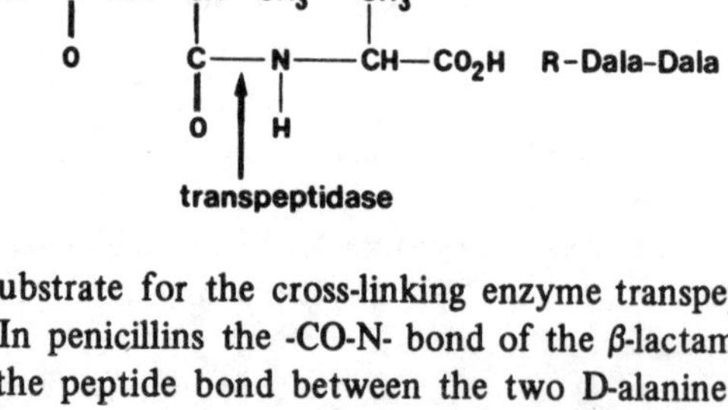

the natural substrate for the cross-linking enzyme transpeptidase (see Figure 3.4). In penicillins the -CO-N- bond of the β-lactam ring is the analogue of the peptide bond between the two D-alanine residues of the natural substrate. During the normal transpeptidation reaction a group in the enzyme reacts with D-alanine-D-alanine, displacing free D-alanine and forming a complex in which a new peptide bond is formed between the enzyme and the penultimate D-alanine. The enzyme is displaced in its turn from this complex by the free amino group at the end of the pentaglycine chain in *Staphylococcus aureus* (Figure 3.5). In the case of the penicillin molecule the enzyme is able

Figure 3.5: Normal Transpeptidation Reaction

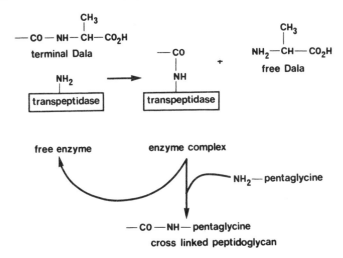

to react with the β-lactam ring forming enzyme-penicillin complex, penicilloyl-transpeptidase, but the reaction cannot proceed further as this complex is stable (Figure 3.6). Penicillin is therefore incorporated

Figure 3.6: Transpeptidation Reaction with Penicillin Forming Stable Complex

as it inactivates the enzyme. The reactive group of the transpeptidase is shown here as an amino group. In one such penicillin-sensitive transpeptidase, the β-lactam ring has been shown to react with, and acylate, the hydroxyl group of a specific serine residue to form the inactive penicilloyl-enzyme complex.

A second type of enzyme, *carboxypeptidase*, analogous to transpeptidase, also occurs in many bacteria. The function of carboxypeptidase is to hydrolyse the D-alanine-D-alanine termini of pentapeptides, releasing free D-alanine without cross-linking. The role of this enzyme seems to vary in different microorganisms. In some bacteria carboxypeptidase competes with transpeptidase thus limiting the extent of cross-linkage. It is suggested that in some strains the action of carboxypeptidase upon a pentapeptide may be a prerequisite for the cross-linking reaction involving the amino side chain of that pentapeptide (Figure 3.7).

Figure 3.7: Carboxypeptidase Action

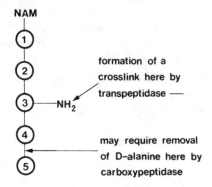

Cephalosporins are a family of antibiotics related to the penicillins in having an identical β-lactam ring but a larger right-hand ring system. The mechanism of action is identical to that of the penicillins. The spectrum of microbial sensitivity to cephalosporins is broad and extends to Gram-negative bacteria.

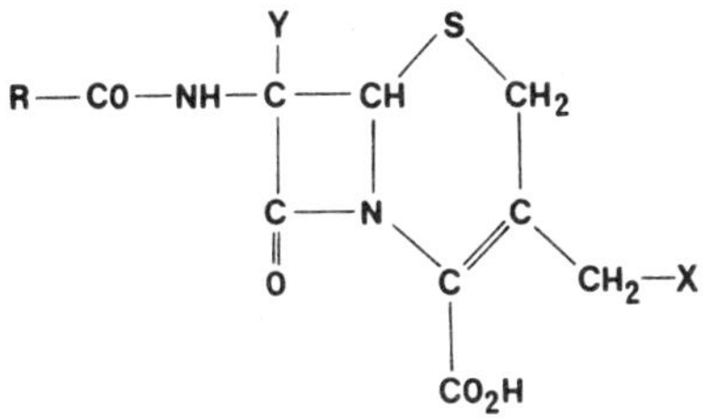

cephalosporin

3.3 Sensitive Bacteria, the Problem of Resistance and the Semisynthetic Pencillins

Beta-lactamases are enzymes that have developed in both Gram-negative and Gram-positive bacteria exposed to penicillins. They are presumed to be mutant variants of the penicillin-binding enzymes, which react with penicillin in the manner described above to produce penicilloyl-enzyme, which is however not a dead-end complex but can be hydrolysed to release free enzyme molecules and inactive penicilloic acid (Figure 3.8). Mutant enzymes of this type therefore destroy penicillin by hydrolysing the β-lactam ring.

Figure 3.8: Hydrolysis of Penicilloyl-enzyme Complex

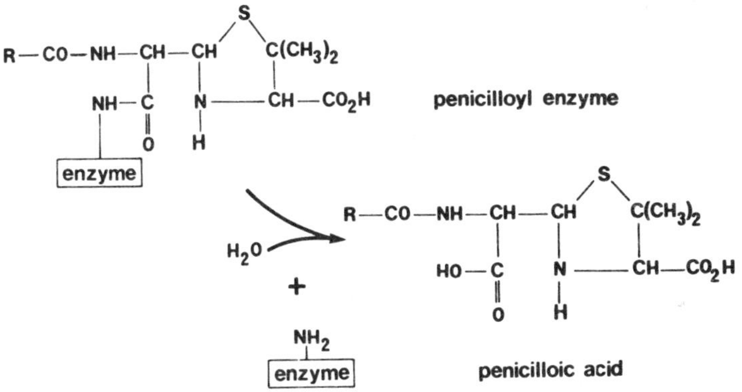

Approximately 70 per cent of hospital isolates of *Staphylococcus aureus* are resistant to penicillin because they produce β-lactamase. Enzymes of this type which hydrolyse penicillin are called *penicillinases*. Analogous mutant enzymes produced by exposure to cephalosporins are terms *cephalosporinases*. Because these two types of β-lactamase do not have the same specificity, penicillinase-producers often remain sensitive to cephalosporins.

Semisynthetic penicillins are produced by treating natural penicillins with a microbial enzyme, *penicillin amidase*, which hydrolytically removes the side chain (Figure 3.9). A wide range of acyl side-chains

Figure 3.9: Production of the Penicillin Nucleus by Penicillin Amidase

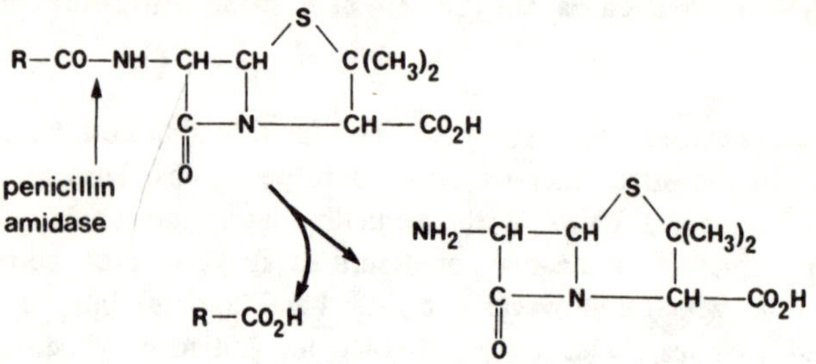

6 - aminopenicillanic acid

have been chemically coupled to the free amino group of the penicillin nucleus providing drugs which have a broader range of antimicrobial activity in some cases and a reduced sensitivity to penicillinase in others. The acyl groups of two natural penicillins (G and V) together with four semisynthetic penicillins are illustrated in Table 3.1. **Ampicillin** is susceptible to staphylococcal β-lactamase but is effective against Gram-negative bacteria. However, the use of ampicillin against infections with enteric bacteria like *Escherichia coli* has resulted in the emergence of resistant strains with a different type of β-lactamase to the staphylococcal enzyme. **Methicillin** was the first semisynthetic member of the group with reduced sensitivity to staphylococcal penicillinase. It has largely been replaced by **cloxacillin** and **oxacillin** which are more stable and have better pharmacological properties. The penicillinase enzyme has such a poor affinity (high K_m) for these molecules that the velocity of hydrolysis is very low at therapeutic concentrations.

Semisynthetic cephalosporins are produced in the same way as

Table 3.1: Side Chains Attached to the Penicillin Nucleus (6-aminopenicillanic acid) in Fermentation and Semisynthetic Penicillins

Penicillin	R group
6 – aminopenicillanic acid	H—
benzyl (Pen G)	
phenoxymethyl (Pen V)	
methicillin	
ampicillin	
oxacillin	
cloxacillin	

semisynthetic penicillins. The stability of cephalosporins to staphylococcal β-lactamase is due to failure of the enzyme to hydrolyse the β-lactam ring. Cephalosporins have two residues that can be varied (see formula above), not only the acyl group R which corresponds to that of the penicillins, but also the acyl group X. In the cephalosporins the residue Y shown in the formula is hydrogen.

Of other β-lactam-type antibiotics, the **cephamycins** are a new class of naturally-occuring antibiotics that resemble the cephalosporins but have a methoxy group in place of hydrogen at Y. These drugs have an extended useful spectrum as the methoxy substituent confers resistance to the β-lactamases of Gram-negative bacteria. Recently, a new type of β-lactam drug has been discovered called **thienamycin** which has a hydroxethyl group in place of the normal amide linked to the β-lactam

$$CH_3-\underset{\underset{OH}{|}}{CH}-CH-\underset{\underset{\underset{\parallel}{C}-N}{|}}{CH} \quad \underset{\underset{C-CO_2H}{\parallel}}{C}\overset{\overset{CH}{\parallel}}{}-S-CH_2-CH_2-NH_2$$

thienamycin

ring, and is stable to β-lactamases. Thienamycin does not have a thiazolidine ring, but does have a sulphur-containing side-chain.

3.4 Pharmacology of Penicillins

3.4.1 Routes of Administration, Absorption and Distribution

Penicillins which are sensitive to stomach acid cannot be given orally if predictable absorption is to occur. Such compounds are given parenterally, usually by intramascular injection. Amongst the *acid-sensitive penicillins* are **benzyl penicillin** (penicillin G), **methicillin, carbencillin, procaine penicillin** and **benzathine penicillin**.

Penicillins which are resistant to the action of stomach acid are often given orally, but may be given parenterally if a rapid onset of action is required. Included in this group are **penicillin V** (phenoxymethyl penicillin), **ampicillin, cloxacillin** and **flucloxacillin**. Despite the fact that benzyl penicillin is largely destroyed by stomach acid in adults it is given orally to those who have less acid than normal: for example in the very young, acid secretion is not fully developed, and in the very old, acid production declines.

When injected intramuscularly, benzyl pencillin is rapidly distributed, but as excretion is rapid, either large doses have to be given, or injections made at frequent intervals, in order that plasma levels are kept adequate. The distribution of benzyl penicillin from the deep intramuscular injection site can be delayed by using procaine penicillin, which is an equimolar complex of procaine and benzyl penicillin in crystalline form; this dissolves and diffuses from the injection site slowly. Benzathine penicillin dissolves and is removed from the injection site even more slowly than procaine penicillin. The procaine in procaine penicillin may have some local anaesthetic activity and decrease the pain of deep intramuscular injection.

Both methicillin and carbenicillin are rapidly excreted following intramuscular injection, but their rate of excretion can be slowed by probenecid. Their duration of action in the body can thus be usefully extended.

When given orally, penicillins are absorbed from both the stomach and duodenum. Any penicillin not absorbed in the small intestine is destroyed by bacterial action in the large intestine. The absorption of orally-given penicillins (penicillin V, ampicillin, cloxacillin and flucloxacillin) can be reduced by the presence of food in the stomach and duodenum, and thus they should usually be taken some time before a meal.

In addition to intramuscular injections, penicillins can be given intravenously, subcutaneously or intrathecally, if the conditions demand it. Preparations of penicillins for topical application are available, but their usefulness is limited to application to wounds and to the eye as they frequently cause rashes when applied to the skin.

Penicillins penetrate into most body tissues, but the concentrations achieved depend on the vascularity and perfusion of the tissue. Penetration into the cerebrospinal fluid is increased when the meninges are inflamed. Penicillins also cross the placental membranes.

The percentage binding of penicillins to plasma proteins varies with the different compounds, but differences in binding do not appear to be major determinants of therapeutic activity, presumably because at the doses used, there is sufficient unbound compound for antibacterial activity. Displacement of penicillins from plasma-protein binding sites, and displacement by penicillins of other drugs from protein binding sites appear not to be a significant cause of interactions.

3.4.2 Metabolism and Excretion

What little metabolism of absorbed penicillins occurs in the body is poorly understood. Only the fact that in kidney failure, the concentration of benzyl penicillin in plasma decreases, while in simultaneous kidney and liver failure there is no decrease in plasma level, suggests that some metabolism may occur in the liver.

While small amounts of penicillin may be lost following metabolism or following secretion into the bile, the great majority of penicillin in the body is excreted by the kidneys in an unchanged form. Up to about ten per cent of penicillin may be filtered through the glomerulus; the vast majority is secreted into the kidney tubules by active transport. Probenecid can block this active transport of penicillins into the lumen of kidney tubules and so it is sometimes used to slow the rate of excretion of penicillins and thus prolong the duration of action of the drug.

In newly-born children, excretion of penicillins is slowed as kidney function is not fully developed. Elimination of penicillins is also delayed in kidney failure, and dosage frequency has to be reduced in order that toxic concentrations do not occur.

A small proportion of cloxacillin is excreted following biliary secretion.

3.4.3 Toxicity

Penicillins do not interact directly with any intermediary metabolic process in man, but adverse reactions do occur in a significant proportion of the population.

Gastro-intestinal disorders such as discomfort, nausea, vomiting and diarrhoea can occur following oral dosing with penicillins. Changes in the gut flora may be responsible for some of these effects. The more serious adverse effects of penicillins are due to either hypersensitivity to penicillins, or to high local concentrations.

The hypersensitivity reactions can take one of two forms. The serum-sickness type of reaction is characterised by rashes, urticaria and fever, and these responses tend to occur some hours or days after administration of the compound. The more serious type of adverse reaction tends to occur immediately following intravenous, intramuscular and occasionally oral administrations of the drug and this is an anaphylactoid reaction which is characterised by cardiovascular collapse, hypotension, weak pulse and shock.

Once hypersensitivity to penicillins has been established, it is usually a sensitivity to all penicillins, and not just to the one to which the first exposure occurred. The occurrence of a serum-sickness response frequently indicates that a more serious anaphylactoid response may occur if a penicillin is used on a subsequent occasion. In persons known to be sensitive to one penicillin, tests for sensitivity to other penicillins can be hazardous, sometimes causing anaphylactic reactions.

Penicillins (when metabolised to penicillenic acid) or substances found mixed with penicillin preparations, can act as antigens, and most people have some antibodies to penicillin although they may be unaware of this. The reason for this might be encounters with penicillins in the general environment, in foods such as milk, and in hospitals where penicillins are found in the atmosphere. The incidence of adverse hypersensitivity reactions is lower in response to a purified form of benzyl penicillin from which some proteins have been removed.

Intrathecal injections of penicillin in high doses have been found to cause loss of consciousness, and in some cases convulsions. Whether these effects are due to the penicillin, or to the high concentrations of sodium or potassium which might have been injected simultaneously, is not clear; nevertheless this form of toxicity is a real danger.

3.5 Pharmacology of Cephalosporins

Cephalothin and **cephaloridine** are cephalosporins which are too poorly absorbed from the gastro-intestinal tract for them to be given orally. They are usually injected intramuscularly, although other parenteral routes can be used. Cephalothin does not cross the blood-brain barrier as readily as does cephaloridine, but penetration of the latter is relatively poor even when the meninges are inflamed. Excretion of all cephalosporins is rapid. Cephalothin is excreted mainly by secretion into the renal tubules; cephaloridine passes into the glomerular filtrate. Metabolism appears not to play a major role in terminating the activity of cephalosporins. **Cephalexin** is absorbed from the gastro-intestinal tract when given orally. It is excreted largely unchanged by the kidneys, and the plasma levels achieved are considerably higher than those of cephalothin or cephaloridine.

Intramuscular injections of cephalosporins can be painful, but cephaloridine causes less pain than cephalothin. Cephalosporins are frequently used as alternatives to penicillins because they only rarely cause a hypersensitivity response in persons who are sensitised to penicillins. Cephalosporins cause sensitisation less frequently than do penicillins. Skin rashes may occasionally occur. Renal damage, possibly to the proximal convoluted tubule, has been associated with high doses (in excess of 8 g/day) of cephaloridine. A lower incidence of renal damage is reported for cephalothin, and cephalexin seems to be free of this adverse effect.

The diuretics frusemide and ethacrynic acid (and probably other diuretics) increase the renal toxicity of cephaloridine. Such combinations should be avoided.

4 INHIBITORS OF NUCLEIC ACID SYNTHESIS

4.1 Introduction

The synthesis of deoxyribonucleic acid (DNA) and ribonucleic acid (RNA) is essential to all living cells. The formation of nucleic acids may be prevented by inhibiting the synthesis of nucleotide bases that are essential components, or by binding to the DNA double helix blocking its template activity in the replication of new DNA molecules, or in the transcription of DNA to RNA. Nucleic acid synthesis can also be blocked by drugs binding to, and interfering with, DNA-dependent RNA polymerase or one of the several enzymes involved in DNA replication.

4.2 Mechanism of Action of Drugs Effective Against Nucleic Acids

4.2.1 Antimetabolites

Prevention of nucleotide synthesis has already been described. Both methotrexate and sulphonamides (see Sections 2.2 and 2.4) block the formation of thymine as one of their actions. **Azaserine** and **diazo-oxo-norleucine** (DON) are antibiotics produced by streptomycetes which inhibit the transfer of the amide nitrogen of glutamine during the

$$
\begin{array}{ccc}
\text{CO}_2\text{H} & \text{CO}_2\text{H} & \text{CO}_2\text{H} \\
| & | & | \\
\text{CH}-\text{NH}_2 & \text{CH}-\text{NH}_2 & \text{CH}-\text{NH}_2 \\
| & | & | \\
\text{CH}_2 & \text{CH}_2 & \text{CH}_2 \\
| & | & | \\
\text{O} & \text{CH}_2 & \text{CH}_2 \\
| & | & | \\
\text{O}=\text{C}-\text{CHN}_2 & \text{O}=\text{C}-\text{NH}_2 & \text{O}=\text{C}-\text{CHN}_2 \\
\text{azaserine} & \text{glutamine} & \text{diazo-oxo-norleucine}
\end{array}
$$

(DON)

synthesis of purine bases, and have been used in tumour therapy (as have several analogues of nucleosides and their bases). Such compounds are not useful in antimicrobial chemotherapy because of their lack of selectivity for bacterial metabolism.

57

4.2.2 Intercalation in DNA

Acridine drugs were used topically as wound disinfectants during the First World War, and are the active ingredients in the yellow ointments used for small burns. Aromatic compounds of this type have been employed as vital stains for eukaryotic tissues because they bind to DNA with changes in their spectral properties. **Ethidium** is used as a

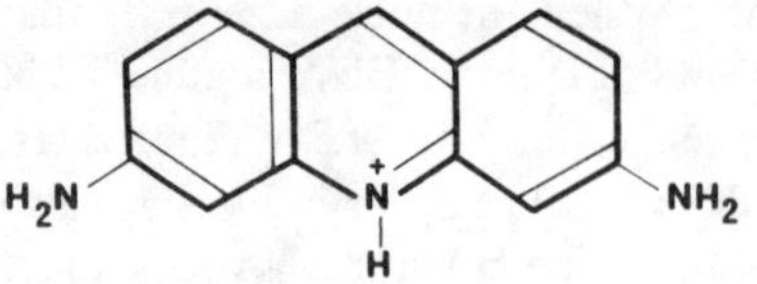

proflavine
(an acridine)

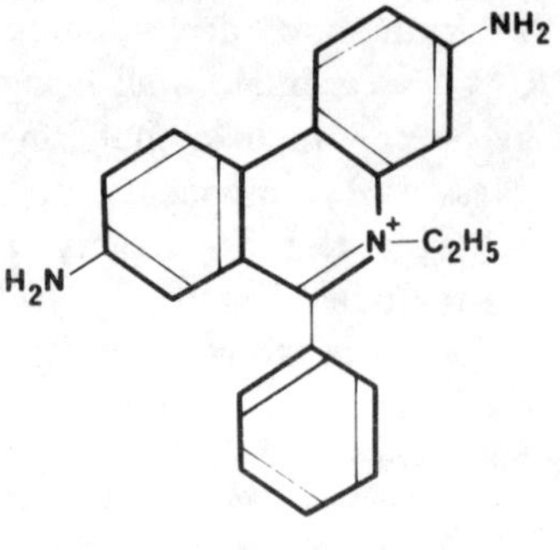

ethidium

reagent for the microassay of DNA because of the change in its fluorescent properties on binding to double-stranded DNA. These dyes are flat molecules which can intercalate between the base pairs of DNA. They destroy the typical X-ray diffraction pattern of double-helical DNA fibres because they disrupt the regularity of the structure. Intercalating drugs also increase the viscosity of DNA solutions by increasing the effective length of the DNA helices, and decrease their buoyant density because the molecular weight of acridines is less than those of the base-pairs that they mimic. By holding the strands of the double helix together, intercalating drugs also increase the thermal stability of DNA.

Acridines were used to produce the 'frameshift' mutants of bacteriophage (virus) T4 that provided evidence for the triplet nature of the genetic code. Figure 4.1 shows how an insertion mutant may be made similar to the wild type by nearby deletion, and how a triple insertion can restore the frame of reading in a hypothetical message coded in triplets.

Treatment of bacterial cells with intercalating drugs inhibits both DNA polymerase and RNA polymerase, although ethidium inhibits RNA synthesis preferentially. Mutagenesis is not often caused in most

Figure 4.1: An Illustration of Frameshift Mutation Induced by
Acridines*

Frameshift Mutations

wild	THE	EGG HAS HEN	DNA BUT THE XXS ARE COX
(+)	THE	EZG GHA SHE	NDN ABU TTH EXX SAR ECO X
(+ −)	THE	EZG GAS HEN	DNA BUT THE XXS ARE COX
(+ + +)	THE	EZG GHZ ASZ	HEN DNA BUT THE XXS ARE COX

⭧ deletions

⭣ insertions of Z

incorrect triplets are underlined

*The message written in triplet words can be understood as long as the frame of reading is maintained. Insertion (or deletion) of a letter makes the messages meaningless to the right of the mutation, as in the second line. The deletion of a letter soon after an insertion makes most of the message intelligible (third line). Similarly three insertions restore the frame of reading (bottom line).

bacteria by such compounds because mutations only appear to occur when crossing-over of chromosomes takes place, which is a rare event in bacteria except during conjugation of Gram-negative bacteria.

Acridine treatment causes the loss of plasmids (extra-chromosomal DNA) in bacteria, the loss of respiratory enzymes specified by mitochondrial DNA, and the loss of the kinetoplast of trypanosomes, all of which involve small units of closed circular DNA.

4.2.3 Actinomycin D and DNA Synthesis

This antibiotic binds to DNA, particularly at guanosine-cytosine base pairs, via the amino and keto groups at the right-hand side of the ring system. One model for the binding, places the two amino acid-containing rings in the narrow groove of DNA. The most important

actinomycin D

biological function of actinomycin D is the selective inhibition of RNA synthesis. The degree of inhibition depends on the ratio of antibiotic to DNA and not on the concentration of enzyme or of nucleotides, indicating that the effect is a direct one on the template. RNA polymerase directed by a polydeoxy (AT) template, which does not bind actinomycin D, is not sensitive to the drug.

The inhibition of RNA polymerase action takes place at concentrations of actinomycin D that do not increase the thermal stability of DNA, which indicates that intercalation is not significant in the process. At higher actinomycin D concentrations the thermal stability of the nucleic acid is increased and the action of DNA polymerase is inhibited. It has been postulated that DNA polymerase action involves interaction between the enzyme and the double helix at the wide groove, while RNA polymerase action takes place in the narrow groove.

Actinomycin D is toxic to eukaryotic cells and is not used for antimicrobial chemotherapy. It has been used to treat Wilms tumour of the kidney in children. The major use of the drug is for experimental purposes, for example to block the synthesis of new RNA after the incorporation of radioactive precursors during pulse-labelling experiments.

4.2.4 *Nalidixic Acid and DNA Synthesis*

Nalidixic acid is a synthetic compound that inhibits the replication of DNA in bacteria, without immediately affecting RNA or protein synthesis in sensitive bacteria. Its toxic effects in animals are limited

nalidixic acid

to inhibiting mitochondrial DNA replication at concentrations that do not affect the nuclear DNA. The existence of bacterial DNA as closed circles of double helix presents topological problems for replication into daughter molecules. Enzymes have been discovered which can nick one of the two strands and facilitate either unwinding a super-coiled loop, or supercoiling a relaxed, closed, circular molecule. Such enzymes, called topoisomerases, include *DNA gyrase*, which requires ATP and relieves the positive supercoiling caused by the progress of the replicating fork around the circular molecule of DNA. This enzyme, which is not present in eukaryotic cells, is the site of action of nalidixic acid. The DNA gyrase molecule is a tetramer of two types of subunit, and nalidixic acid binds to the larger 'A' subunit. Mutations to nalidixic acid resistance occur in the *nal-A* gene of *E. coli* and result in A protein that does not bind the inhibitor.

4.2.5 *Rifampicin and RNA Polymerase*

Rifampicin is a semisynthetic derivative of the complex **rifamycin** antibiotic nucleus. The drug has no affinity for nucleic acids but binds avidly to DNA-dependent RNA polymerase. The enzyme from *E. coli* is 50 per cent inhibited by 2×10^{-8}M rifampicin, one-tenth of the concentration of actinomycin D required to achieve the same effect with *in vitro* RNA synthesis. DNA polymerase I is not inhibited by 10,000 times this concentration of the drug. RNA polymerase is completely inhibited when one molecule of rifampicin binds per enzyme molecule.

The synthesis of RNA by DNA-dependent RNA polymerase from rifampicin-resistant strains of bacteria is resistant to rifampicin *in vitro*.

61

The bacterial RNA polymerase enzyme comprises five polypeptide chains of four types, two α, one β, one β^1 and one σ subunit. A β subunit from a resistant strain is sufficient to confer rifampicin resistance on the reconstituted enzyme irrespective of the source of the other subunits, which shows that rifampicin binds to the β polypeptide in order to inhibit the enzyme.

4.3 Sensitive Bacteria and Acquired Resistance

Intercalating drugs are of many types and some of them (e.g. **acriflavine** and **proflavine**) are used in general antiseptic creams and jellies for topical application. Many of them are too toxic for systemic use, but an acridine drug (**mepacrine**) has been used against the plasmodium parasite of malaria, and ethidium is useful against trypanosomes, but neither of these are bacteria and so are outside the scope of this book. Actinomycin D is of no use in the chemotherapy of infections.

Nalidixic acid is a synthetic drug which is excreted, together with inactive metabolites, in the urine. It is bactericidal to most of the Gram-negative bacteria that cause the common urinary tract infections, except for pseudomonas strains. Resistance to nalidixic acid develops readily but is apparently not transferable from one organism to another.

Rifampicin inhibits the growth of most Gram-positive together with some Gram-negative strains such as *Escherichia coli*, *Pseudomonas*, *Proteus* and *Klebsiella*. It is the drug of choice for the prophylaxis of meningococcal infections in the persons at risk because they have been in contact with affected individuals, and it is also used in the treatment of pulmonary tuberculosis.

Mycobacteria and other bacteria may develop a resistance to rifampicin by a rapid one-step process, and therefore the drug should not be used in isolation. Resistance develops by mutation of the DNA-dependent RNA polymerase, and resistant strains have appeared in as short a time as two days of treatment.

4.4 Pharmacology of Nalidixic Acid and Rifampicin

These are the only antibiotics which directly inhibit nucleic acid synthesis and which are used therapeutically.

Nalidixic acid is given orally because it is absorbed from the gastro-intestinal tract. A large proportion of the drug is bound to plasma proteins and penetration into the tissues is poor. The plasma half-life is about 1.5 hours, and the major route of excretion of both parent compound and metabolites is via the kidney. The metabolites of nalidixic acid are bacteriologically inactive, but sufficient unmetabolised compound is excreted into the urine to make nalidixic acid useful in the treatment of urinary tract infections.

The toxic effects of nalidixic acid include nausea and vomiting, skin rashes and photosensitivity reactions, central nervous system effects such as dizziness and visual disorders, and, in persons prone to them, convulsions. Nalidixic acid can displace warfarin from protein binding sites and can prolong the clotting time, or cause haemorrhage, if these drugs are given together.

Rifampicin is well absorbed when given orally, and as it is excreted slowly, prolonged therapeutically-effective plasma concentrations are achieved. Rifampicin penetrates all body compartments and may cause sweat, tears and urine to be coloured pink. The major metabolite of rifampicin, formed in the liver, is as active as the parent compound. Excretion is mainly by secretion into the bile, and as some reabsorption occurs, enterohepatic circulation may contribute to the long half-life of the drug in the body. Rifampicin accumulates in the body during liver failure or during biliary-duct obstruction, but some urinary excretion can take place when plasma concentrations are high.

Rifampicin is used mainly for the treatment of tuberculosis and leprosy. All populations of bacteria contain some individuals which are resistant to the action of rifampicin, and therefore it must always be used in combination with another drug to which the bacteria are sensitive (see Section 5.3).

Adverse reactions to rifampicin are relatively rare, and effects on the liver are frequently reversible if the drug is discontinued. The side effects most frequently encountered are rashes, diarrhoea, nausea, vomiting and lethargy following large doses of the drug.

5 INHIBITORS OF PROTEIN SYNTHESIS

5.1 Introduction

Inhibitors of RNA synthesis cause secondary cessation of protein synthesis, but those compounds which are direct inhibitors of the assembly of polypeptides have a specific and rapid effect on the accumulation of protein in growing cultures. Thus addition of an antibiotic such as chloramphenicol to a rapidly-dividing culture of bacteria will stop protein synthesis within a few minutes, while RNA and DNA synthesis continue unaltered for an hour or more.

There are many inhibitors of protein synthesis and most are bacteristatic, preventing the growth of sensitive bacteria. One important group, the aminoglycosides, including streptomycin, gentamicin, neomycin and kanamycin, are bactericidal for reasons which remain unknown. The killing action of aminoglycosides has been attributed to the multiple effects of these antibiotics on membranes, RNA metabolism and misreading of codons, but it is now clear that the effect of members of this group depends on their concentration and differs from drug to drug.

5.2 Mechanisms of Action

5.2.1 Outline of Protein Synthesis in Bacteria

The linking together of amino acids in specific sequences by means of peptide bonds takes place on the ribosomes of all cells. The instructions for the sequence are transcribed from the genes of the chromosomal DNA into a single strand of *messenger RNA* (mRNA). A coding unit (codon) of mRNA comprises three nucleotide bases which represent a particular amino acid. Ribosomes in the process of synthesising protein are held together in chains (polysomes) by the mRNA which specifies the sequence of the protein. Ribosomes that are not engaged in protein synthesis are dissociated into their constitutent subunits (in bacteria the 30S and 50S components, equivalent to the 40S and 60S subunits of eukaryotic cells). Protein synthesis begins with the assembly of the ribosome together with a mRNA molecule, and the *initiator transfer RNA* (tRNA) coupled to *N*-formyl methionine (fmet), which binds to the codon AUG in the mRNA molecule (Figure 5.1). The mRNA molecule associates with the smaller, 30S, ribosomal subunit, which recognises a purine-rich sequence to the left of AUG (not shown in Figure 5.1), i.e. towards the $5'$ end of the mRNA strand. Three initiation protein factors are needed (Figure 5.1), of which IF_2 acts as an aggregate with the energy-supplying nucleotide GTP

Figure 5.1: The Formation of the Initiation Complex for Protein Biosynthesis*

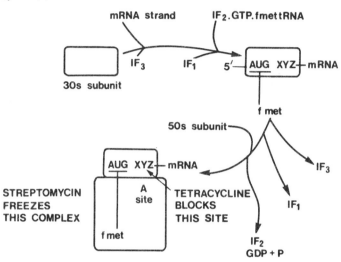

*The assembly involves the 30S and 50S subunits of the ribosome, three protein factors, IF_1, IF_2 and IF_3, messenger RNA, formyl-methionyl-tRNA$_F$ and GTP. Tetracycline acts by blocking access to the aminoacyl-tRNA (A-site) while streptomycin prevents the process of synthesis progressing any farther.

and the fmet-tRNA$_F$ molecule. The initiation complex is completed by the addition to the assembly of the 50S ribosomal subunit and the release of the three protein factors, IF_2 being released associated with GDP. At this stage the peptidyl binding site (P-site) of the 50S component is occupied by formyl-methionyl-tRNA$_F$ and the aminoacyl binding site (A-site) is vacant. The codon in mRNA opposite to the A-site (XYZ in Figure 5.1) will select which aminoacyl-tRNA can enter the A-site, and therefore which amino acid will be linked to formyl methionine by a peptide bond. **Tetracycline** and **streptomycin** are

Inhibitors of Protein Synthesis

two antibiotics that affect protein synthesis at this stage of the ribosomal cycle.

The entry of aminoacyl-tRNA (Figure 5.2) into the ribosome resembles that of fmet-tRNA in that a complex is formed between

Figure 5.2: Introduction of Aminoacyl-tRNA and the Synthesis of a Peptide Bond*

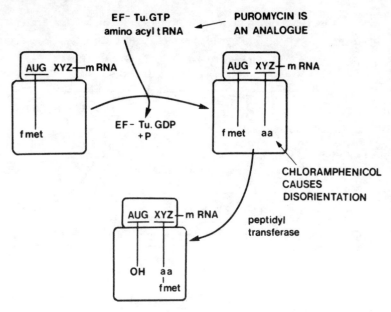

*The binding of aminoacyl-tRNA to the A-site requires elongation factor EF-Tu and GTP to be complexed with the aminoacyl-tRNA molecule. Peptidyl transferase makes a new peptide bond by transfering formyl methionine from its tRNA molecule and combining it with the aminoacyl-tRNA at the A-site. Puromycin acts as an analogue of aminoacyl-tRNA and enters the A-site in its place. Chloramphenicol disorientates the aminoacyl-tRNA molecule so that peptide bond synthesis is unsuccessful.

the aminoacyl-tRNA, a protein factor (EF-Tu) and the nucleotide GTP. Elongation factor Tu is released associated with GDP derived by hydrolysis of this bound GTP. Elongation factor Ts (EF-Ts), which is involved in converting EF-Tu:GDP back to EF-Tu:GTP by replacing the nucleoside triphosphate, is not shown in Figure 5.2. The presence of aminoacyl-tRNA in the A-site completes the requirements for the formation of a peptide bond. The enzyme peptidyl transferase is a constituent of the 50S ribosomal subunit and transfers fmet (or the growing peptide in later cycles of synthesis of the protein) from the tRNA molecule in the P-site to the amino acid (as in Figure 5.2) in the A-site, so forming a new peptide bond. Two antibiotics which affect this stage of the synthetic process are **puromycin** and **chloramphenicol**.

The last stage of each cycle of synthesis is translocation (Figure 5.3), the movement of the ribosome relative to the mRNA molecule.

Figure 5.3: Translocation of mRNA Relative to the Ribosome*

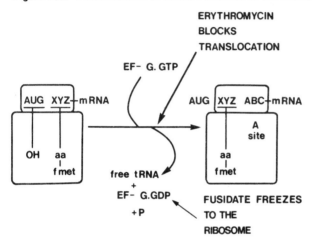

*This produces a vacant A-site ready to accept a new aminoacyl-tRNA molecule. GTP linked to elongation factor G is hydrolysed during this process and EF-G combined with GDP is released. Erythromycin inhibits the translocation while fusidate prevents the release of the complex of EF-G and GDP from the ribosome after translocation has occurred.

In this process the tRNA molecule which has just lost formyl methionine (or a peptide in later cycles), and now has a free -OH group, leaves the P-site of the ribosome. The tRNA bearing the new peptide enters the P-site leaving the A-site vacant and a new codon (ABC in Figure 5.3) appears opposite site A and selects the next aminoacyl-tRNA that can enter site A. The process of translocation requires energy, which is supplied as GTP complexed with factor G. The antibiotic **erythromycin**, which belongs to the macrolide group, and the steroid-like antibiotic, **fusidate**, both interfere with ribosome translocation. The vacant ribosomal site A may now accept a new aminoacyl-tRNA, and the cycle of events depicted in Figures 5.2 and 5.3 is repeated for each new amino acid incorporated into the polypeptide chain. When the peptide is complete a 'stop' codon enters the ribosome opposite to the A-site. A termination factor probably binds to this codon and catalyses the hydrolytic removal of the polypeptide from the tRNA molecule in the P-site. The nascent polypeptide is thus released and the ribosome dissociates into its components which are available for reassembly into a new initiation complex.

5.2.2 Puromycin, a Non-selective Inhibitor

Although the process of protein synthesis is basically the same in bacteria and eukaryotes, there are differences in the detail which permit some antibiotics to act selectively against bacteria. **Puromycin** is an analogue of aminoacyl-tRNA; the latter is sufficiently uniform in all cells for the antibiotic to be inhibitory in all protein-synthesising systems. The nucleoside of puromycin is an analogue of the adenosine by which amino acids are joined to all tRNA molecules, and comprises

aminoacyl tRNA   puromycin

dimethyl-adenine linked to 3-amino-ribose. The amino acid analogue in puromycin (a tyrosine derivative) is linked to the nucleoside by a peptide bond whereas the amino acid in aminoacyl-tRNA is joined to ribose by an ester bond. When puromycin enters the A-site of a ribosome it can accept the polypeptide chain, which is linked to the free amino group of the antibiotic by *peptidyl transferase*. However, the process of polypeptide chain extension cannot proceed any further as the polypeptidyl puromycin detaches itself from the ribosome. Puromycin is therefore covalently linked to the carboxyl ends of prematurely-terminated polypeptides. The puromycin reaction requires no energy, which provides some evidence that the normal formation of peptide bonds, to which the reaction is analogous, also needs no energy.

As it is non-selective, puromycin finds no clinical use, but it is valuable experimentally and as a tool for investigating the mechanism of other antibiotics.

5.2.3 Streptomycin and Other Aminoglycosides

Streptomycin was discovered during a survey to find an antibiotic effective against Gram-negative bacteria to complement the penicillins, which had proved to be so valuable against Gram-positive organisms. Streptomycin, which is also effective against some Gram-positive bacteria, comprises three carbohydrate ring structures linked by glycosidic bonds and is strongly basic. Other useful aminoglycosides are **gentamycin, kanamycin** and **neomycin**.

streptomycin

Streptomycin, as the first member of the family to be discovered, has been investigated the most and it has a multiplicity of effects upon sensitive microorganisms. Some membrane damage is caused, as evidenced by the loss of ions and low molecular weight metabolites, and respiration is inhibited in cells with aerobic metabolism. It also causes increased RNA synthesis and mistranslation of the genetic message into proteins, and inhibits protein synthesis. However, highly streptomycin-resistant mutants of *E. coli* are different to the sensitive parent strain in one ribosomal protein, so that the critical effects of streptomycin are those concerned with ribosome function.

Whole cells of *E. coli* exposed to low concentrations of streptomycin (2 μg/ml) synthesise protein which contains miscoding errors, whereas cells treated with the higher concentrations (20 μg/ml or more) cease to make protein.

The synthesis of protein *in vivo* with components derived from streptomycin-sensitive or streptomycin-resistant bacteria reveals that the antibiotic binds to the 30S subunit. Protein synthesis is sensitive to streptomycin in a cell-free system in which the 30S subunit is from streptomycin-sensitive cells, and all other components are from streptomycin-resistant cells. The converse is true when the 30S subunit is the only component from streptomycin-resistant bacteria. The 30S ribosomal particle can be separated into component RNA and protein molecules, from which the active subunit may be reassembled. The streptomycin-binding capacity resides in one protein, originally known as the P10 protein, which is the site of mutation in streptomycin-resistant cells of *E. coli*. Ribosomal proteins have now been renumbered by a system which indicates whether they occur in the large (L) or small (S) subunit. Protein P10 is known as protein S12 according to this scheme.

Experiments to determine the mode of action of streptomycin *in vitro* have produced conflicting results. During protein synthesis the ribosome population includes free 30S and 50S subunits, and polysomes comprising 70S ribosomes held together by the mRNA that they are translating into protein molecules. When streptomycin is added to such a system the number of ribosomes attached to each mRNA molecule falls to one; the polysomes become 'streptomycin monosomes'. This is interpreted as the binding of streptomycin to ribosomes in the initiation-complex state followed by their failure to traverse the mRNA and to synthesise proteins. Ribosomes already in the process of synthesising a polypeptide molecule continue until it is completed, and then become detached from the mRNA in the normal way. The blocked initiation-complex ribosome is immobile and denies access of other ribosomes to the mRNA strand. The conflicting results of protein synthesis experiments *in vitro* have been attributed to two alternative effects of the binding of streptomycin to the 30S subunit. A ribosome that is traversing a messenger RNA molecule and synthesising polypeptide will continue to do so (albeit at a slower rate, and with mistakes in amino acid incorporation) if streptomycin binds to the S12 protein. Such will be the case if isolated polysomes are allowed to complete their polypeptide chains, and 'streptomycin monosomes' will be the end result. However, if 30S ribosomes bind streptomycin before, or during, the formation of the initiation complex, initiation is not successful and protein synthesis does not occur. It is this blockage of protein synthesis rather than any of the other effects of streptomycin that is considered to be bactericidal. However, other antibiotics that inhibit protein synthesis are only bacteristatic, and cells washed free of antibiotic are still viable. In the case of the aminoglycosides it may be that the binding of the drug to ribosomes is so strong that the ribosomes remain inhibited long after the cells have been washed free of unbound antibiotic.

5.2.4 The Tetracyclines

These antibiotics are a very similar group of compounds produced by *Streptomyces* strains. They vary somewhat in their pharmacology (see 5.4.2) but resemble one another closely in antimicrobial activity, and when resistance arises it is usually to all tetracyclines. The first member to be discovered was **chlortetracycline** which has chlorine substituted at position 7 of the nucleus. Tetracycline is now usually

tetracycline

made semisynthetically from chlortetracycline although it was originally isolated as a fermentation product. **Oxytetracycline**, another commonly-used member of the group, has a hydroxyl substituted at position 5. At the plasma concentrations that can be achieved in

71

bacterial chemotherapy, the drugs are bacteristatic although they are bactericidal at the higher concentrations that can be achieved *in vitro*.

The mechanism of tetracycline action has been variously proposed to be by inhibition of transport across membranes, of the metabolism of glucose and of oxidative phosphorylation. However, as early as 1953 tetracyclines were known to inhibit protein synthesis.

Incorporation of phenylalanine directed by poly (U) in a cell-free system is less sensitive to tetracyclines if the 30S ribosome particle is derived from a tetracycline-resistant strain. The 50S subunit plays no part in the inhibition. The A-site of the ribosome is the site at which tetracyclines produce their effect after having bound to the 30S par-ticle. When ribosomes are mixed with the synthetic nucleotide poly (A), tetracyclines inhibit the binding of lysyl-tRNA but not the corres-ponding peptide derivative polylysyl-tRNA which may be expected to bind to the P-site. Protein synthesis *in vitro* with bacterial (70S) or eukaryotic (80S) ribosomes is inhibited equally by tetracyclines. How-ever, these antibiotics are actively accumulated by both Gram-positive and Gram-negative bacteria by an energy-dependent process, while eukaryotic cells do not concentrate tetracyclines and so are little affected by the usual therapeutic doses. Treatment with drugs that alter membrane permeability may however facilitate tetracycline entry, and so cause protein synthesis to be inhibited in mammalian cells by tetracycline doses that would otherwise not be anti-anabolic (see 5.4.2).

5.2.5 Chloramphenicol

Although originally isolated from a streptomycete, **chloramphenicol** is one of the simplest of antibiotics and is cheaply manufactured by chemical synthesis rather than by fermentation. The molecule exists in

chloramphenicol

four isomeric forms of which only D-threo-chloramphenicol is effec-tive. The drug stops protein synthesis rapidly without affecting DNA or RNA synthesis, and it is specific in not binding to eukaryotic 80S ribosomes. However, the effectiveness of chloramphenicol in blocking protein synthesis on 70S ribosomes does cause some problems in rapidly-dividing tissues, because the reproduction of mitochondria requires some proteins, coded for by mitochondrial DNA, to be synthesised on mitochondrial rather than on cytoplasmic ribosomes. Several features of mitochondria suggest that they may have been derived from much-reduced symbiotic bacteria, and one of these is that mitochondrial ribosomes are of the 70S type, and are therefore sensitive to chloramphenicol. The subunit to which the drug binds is the 50S particle. The binding of aminoacyl-tRNA and mRNA to the 30S subunit is unaffected. Chloramphenicol binds relatively weakly to the 50S subunit of the ribosome, and the binding is not blocked by antibiotics such as tetracycline and streptomycin, which attach themselves to the 30S component. It inhibits the formation of peptide bonds without blocking the translocation of mRNA relative to the ribosome, although the velocity of translocation is reduced, as shown by the reduced rate of dissociation of ribosomal subunits from polysomes.

Evidence that chloramphenicol prevents peptide-bond formation is provided by the inhibition of the puromycin reaction in intact bacteria, and also of polylysyl puromycin and similar compounds *in vitro*. The mechanism may be to disrupt the normal positioning of aminoacyl-tRNA so that peptidyl transferase cannot operate. Thus degraded aminoacyl-tRNA, from which much of the polynucleotide has been removed by ribonuclease leaving the portion close to the amino acid intact, binds to the ribosomes in the absence of mRNA. This binding is specifically prevented by the microbiologically active D-threo-chlorampehnicol, but not by the other three isomers. Chloramphenicol therefore seems to affect the orientation of that part of aminoacyl-tRNA which is close to the amino acid residue.

5.2.6 Fusidic Acid

This antibiotic is unusual in having a structure that resembles the steroids. It is usually employed as the more soluble sodium salt in the UK but it is not used at present in the USA. Fusidate is effective against Gram-positive but not Gram-negative bacteria, although it will inhibit protein synthesis in cell-free preparations of Gram-negative bacteria. Translocation is inhibited and the release of tRNA from the P-site, together with conformational changes in the ribosomes that normally accompany translocation, may both be prevented. The direct effect of fusidate seems to be to bind to the complex of elongation factor EF-G and GDP, released as translocation occurs. Thus the hydrolysis of excess GTP by EF-G in the presence of functional ribosomes is halted by fusidate, and EF-G complexed with GDP accumulates. Therefore EF-G is rendered unavailable for further cycles of ribosome translocation. EF-G from fusidate-resistant bacteria fails to bind fusidate in this way, and if EF-G from such a source is used in an *in vitro* system, protein synthesis is uninhibited by fusidate.

5.2.7 Erythromycin

The macrolide antibiotics have structures that include large (12-16 atoms) lactone ring systems, and carbohydrate side chains that include amino sugars. The macrolide group includes **spiramycin, oleandomycin, lankamycin** and **carbomycin**, but the most important clinically is **erythromycin**. Macrolides inhibit protein synthesis in bacteria and also in cell-free systems derived from bacteria. Erythromycin binds to the 50S ribosomal subunit of sensitive bacteria but not to that of resistant mutants. However most of the tests for inhibitor activity against ribosome function are negative and it is believed that the effect of erythromycin may be upon the translocation of ribosome relative to mRNA, although it may be inhibitory to the peptidyl transferase enzyme.

5.3 Sensitive Bacteria and Resistance

The aminoglycoside family is still widely used, the main limiting factors being its pharmacokinetics and toxicity. **Streptomycin** together with penicillin is used for streptococcal infections, and either gentamycin or kanamycin may be employed in the treatment of many gram-negative bacilli such as *Enterobacter*, *Proteus*, *Pseudomonas* and *Shigella*. Streptomycin together with other chemotherapeutic agents is the treatment of choice for tuberculosis caused by *Mycobacterium tuberculosis*.

Resistance to aminoglycosides often involves the mutation of the binding site in the ribosome. Transferable drug resistance also occurs and depends on introduction of a gene for an aminoglycoside-inactivating enzyme into a sensitive strain by way of a plasmid. Inactivation reactions catalysed by such enzymes involve ATP and the aminoglycoside is either phosphorylated or adenylated.

The **tetracyclines** are all very similar in their antimicrobial spectrum, which is wide. A tetracycline is the drug of first choice for relatively microorganisms such as the cholera vibrio, and a number of peculiar microorganisms such as mycoplasmas, chlamydia (including trachoma) and rickettsia (including typhus). Tetracyclines are also drugs of second choice for a wide range of Gram-positive and Gram-negative bacteria. As antibiotics of broad spectrum that can be administered orally they are often prescribed for low-grade infections in outpatients.

Resistance to tetracyclines seems to develop slowly and in stages and involves changes in membrane permeability. Some bacteria actively accumulate tetracyclines, and plasmid-mediated resistance in *E. coli* seems to depend on the synthesis of some component necessary for exclusion of tetracyclines from the cell. Members of the family of drugs are so similar that cross-resistance is complete.

Chloramphenicol is a moderately wide-range antibiotic effective against Gram-positive and Gram-negative bacteria as well as myco-plasmas, rickettsia and chlamydia. However, because of its toxicity its use is limited to severe infections that do not respond to other anti-biotics. It is the drug of first choice for typhoid fever (caused by *Salmonella typhi*) which sensitivity testing shows is not resistant to chloramphenicol. It is the second choice to tetracyclines for treatment of mycoplasmas, rickettsia and chlamydia. Resistance to chloramphenicol occurs in both Gram-negative and Gram-positive bacteria, and is transmissible in the Gram-negative enteric genera in which a plasmid specifies an acyltransferase that catalyses the transfer of an acetyl group from acetyl-CoA to chloramphenicol, thus inactivating the drug.

Erythromycin is most effective against the Gram-positive cocci and many of the Gram-positive rods. Gram-negative bacteria only achieve about one per cent of the intracellular concentration of the drug that can be obtained with Gram-positive strains. Its main use is against staphylococcal, streptococcal and pneumococcal infections, but it is the drug of first choice for treatment of meningitis caused by *Flavobacterium meningosepticum*, a Gram-negative rod. It is a drug of second choice for most Gram-positive and some Gram-negative bacteria as well as some mycoplasmas and chlamydia.

5.4 Pharmacology and Toxicity

5.4.1 *Streptomycin and Other Aminoglycosides*

Pharmacology. Aminoglycoside antibiotics are not absorbed from the gut, but are lost in the faeces. Therefore they are not given orally unless an effect in the lumen of the gut is required. Parenteral administration by deep intramuscular, subcutaneous or intrathecal injection may be used, but intravenous injection is the most rapid means of achieving therapeutically-effective concentrations. Some absorption can occur following topical application to open wounds, and this may be sufficient to cause toxic side effects. **Neomycin** can be given by inhalation of an aerosol for the treatment of respiratory infections. The lipid-solubility of aminoglycoside antibiotics is low, as may be surmised from their poor absorption from the gastro-intestinal tract. Once in the blood-stream they do not readily cross most membranes, but tend to be restricted to extracellular fluid spaces. Entry into the cerebrospinal fluid is poor except when the meninges are inflamed. **Streptomycin** equilibrates with foetal blood and thus apparently crosses the placenta. Plasma half-life of aminoglycosides has been estimated to be between two and four hours, but this can increase dramatically in the event of kidney failure.

Aminoglycosides do not undergo significant metabolism in the body — presumably because they do not readily enter the cells. Glomerular filtration occurs and the major route of excretion is via the urine. During kidney failure, dosage frequency has to be reduced so that toxic concentrations do not occur in the plasma. Excretion of aminoglycosides is also reduced in infants when kidney function is not yet fully developed. Biliary excretion is not believed to be significant in removal of aminoglycoside antibiotics from the body.

Toxicity. Hypersensitivity reactions to aminoglycosides are independent of the dose given, and occur in about five per cent of persons. They are characterised by rashes and fever, though more rarely blood dyscrasias or anaphylactoid reactions may occur. Dermatitis can occur when the drugs are applied topically. Intramuscular injections of streptomycin are painful, and therefore procaine is sometimes injected with the antibiotic as a local anaesthetic.

Damage to the 8th cranial nerve (*ototoxicity*) can occur during or following treatment with the aminoglycosides, and is proportional to the dose and duration of exposure. This ototoxicity may include either, or both, labyrinthine and auditory effects. When labyrinthine damage occurs, the patient experiences headaches, dizziness, vertigo and loss of balance when walking, or when making sudden movements. Some compensation and return of vestibular function occur when the drug is withdrawn, but this can take several months. Auditory damage may sometimes result in total deafness, the early signs of which are loss of perception of high frequencies, often accompanied by a ringing in the ears. The drug should be withdrawn at this stage to avoid further deterioration in hearing. Toxic effects are more likely to occur when high doses of aminoglycosides accumulate during poor kidney function, for example in the aged.

When **streptomycin** has been administered as a powder or a strong solution to the peritoneal cavity following surgery, potentially-fatal respiratory parlaysis may occur. This is due to the neuromuscular-blocking property of streptomycin, which is similar to that of (+)-tubocurarine. Persons suffering from myaesthenia gravis, or those who have recently received a muscle relaxant are most likely to have such a response to streptomycin.

Neomycin is the most toxic of the aminoglycosides as far as damage to the auditory nerve is concerned. Its pharmacokinetics are very similar to those of streptomycin if it is given intramuscularly, but parenteral use is now extremely rare. Neomycin is used topically for treatment of superficial infections, and it is frequently combined with bacitracin or polymyxin when used in this way. Neomycin has been given orally to suppress the gut flora and to sterilise the bowel. Ototoxicity can occur if such treatment is prolonged. Paralysis of respiration has been noted when neomycin has been applied near the diaphragm, and inhalation of neomycin as an aerosol for treatment of lung infections can also cause deafness.

Kanamycin is less ototoxic than neomycin; thus under carefully-controlled conditions it can be given intramuscularly to treat septicaemia.

Gentamycin is the most potent of the aminoglycoside antibiotics, and shares many properties with streptomycin. It has to be given parenterally and excretion is through the kidneys. The toxicity of gentamycin is directed mainly towards labyrinthine damage, and it is more liable to cause such effects than kanamycin. The nephrotoxicity of gentamycin is not firmly established. Ototoxicity occurs more frequently in persons with poor renal function, presumably as a result of reduced clearance from the body.

5.4.2 Tetracyclines

Tetracyclines are basic compounds which are only sparingly soluble in water, but which form readily-soluble sodium salts. Whereas the bases and salts are stable when dry, biological activity is lost in solution especially at alkaline pH values. Suitably-prepared tetracyclines are available for administration orally, parenterally, and topically to the eye.

Absorption. Tetracycline absorption from the gastro-intestinal tract is not the same in all individuals, but absorption from the stomach and upper small intestine often produces therapeutically-effective plasma concentrations. **Doxycycline** and **minocycline** are well-absorbed from the gastro-intestinal tract, and the presence of food does not interfere with their absorption. **Chlortetracycline, oxytetracycline, tetrcycline** and **methacycline** are not so well-absorbed, and the presence of food can further decrease absorption. The reason for irregular absorption of tetracyclines from the gastro-intestinal tract is not well-understood, but it is suggested that chelation of calcium ions (e.g. from milk) may cause precipitation of insoluble complexes in the lumen of the gut. Phosphates may decrease the availability of free calcium in the lumen of the gut and so facilitate absorption of tetracyclines. Iron salts and antacids such as aluminium hydroxide and sodium bicarbonate, also decrease absorption. To avoid uncertainty of gastro-intestinal absorption tetracyclines are often given intravenously.

Distribution. Tetracyclines are bound in varying degrees to plasma proteins, but plasma concentrations of unbound drug are adequate to ensure passage into most of the tissues. Tetracyclines are absorbed by the liver, passed into the bile and are excreted into the gastrointestinal tract, from which some reabsorption may occur. Tetracycline concentration in cerebrospinal fluid may rise to about 25 per cent of the plasma value. Because of their irritant properties, tetracyclines must not be given intrathecally. Tetracyclines are deposited in developing teeth and bones, presumably as calcium complexes and discoloration of teeth may persist months after exposure to tetracyclines. Therefore children younger than eight years are not treated with these antibiotics. Tetracyclines also cross the placenta and should not be given to pregnant women.

Metabolism. The metabolism of all tetracyclines occurs in the liver, but to a degree that depends on the individual drug. The microsomal enzymes appear to have a role in metabolism, as the half-life in plasma of doxycycline can be shortened by induction of these microsomal enzymes, e.g. with phenobarbitone.

Excretion. Tetracyclines are excreted by the kidney in the glomerular filtrate, and in the faeces after secretion in the bile, although some faecal tetracycline may result from failure to absorb orally-administered doses. Changes in these two routes of excretion affect the duration of action of tetracyclines. If glomerular filtration is low, or poor liver function decreases the amount secreted into the bile, then plasma half-lives will be prolonged. Normally, some of the tetracycline secreted into the bile is reabsorbed in the intestine. The significance of entero-hepatic circulation to the therapeutic action of tetracyclines is not known.

Toxicity. Skin rashes and urticaria can occur in sensitive individuals after treatment with any of the tetracyclines, and cross-sensitivity to tetracyclines other than the one which first precipitated the sensitivity reaction is general. Allergic reactions of the anaphylactoid type can occur.

Following oral intake, gastro-intestinal disturbances such as nausea, vomiting and pain can occur, and are usually dose-dependent. Diarrhoea may also occur due to irritation of the mucosa, in which case the stools are free of blood and leucocytes. An alternative cause of diarrhoea is superinfection of the gut following the suppression of the normal bacterial population, in which case blood and leucocytes are found in the watery stools. In this life-threatening condition it is necessary to withdraw the tetracycline, identify the microorganism responsible for the superinfection and suppress it with an alternative antibiotic, and then restore fluid and ionic balance by intravenous fluid infusion.

Tetracyclines have an anti-anabolic action; they facilitate protein catabolism and reduce protein synthesis. They should not be used on malnourished patients in whom weight loss and negative nitrogen balance may be caused. The exact relationship has not been established between this action and the incidence of liver damage and worsening of kidney function following tetracycline administration, but a causal relationship seems likely. Liver and kidney damage usually occurs when doses of tetracyclines exceed 2 g per day. Liver damage is characterised by fatty deposits, and results in jaundice and a high nitrogen excretion. The livers of pregnant women appear to be especially sensitive, which is a good reason for avoiding the use of tetracyclines in pregnancy. There is no evidence for teratogenic effects of tetracyclines, but unsightly pigmentation of tooth enamel may occur due to deposition of the drugs in developing hard tissues. The anti-anabolic action of tetracyclines may exacerbate kidney failure especially if the kidneys themselves are infected.

5.4.3 Chloramphenicol

Routes of Administration and Absorption. Chloramphenicol is well absorbed from the gastro-intestinal tract when it is present as 'microfined' particles. More commonly, **chloramphenicol palmitate** is given orally, and **chloramphenicol sodium succinate** is used for intravenous injection. Chloramphenicol palmitate is hydrolysed in the gut before absorption takes place. Chloramphenicol penetrates all tissue compartments, and effective concentrations are found in the cerebrospinal fluid.

Metabolism and Excretion. Chloramphenicol is conjugated in the liver to form a glucuronide which is not bactericidal. The metabolism of chloramphenicol by this mechanism can be speeded if microsomal liver enzymes have been induced by drugs such as phenobarbitone or phenytoin.

Excretion of chloramphenicol and the glucuronide conjugate is principally via the kidney, the parent compound being filtered by the glomerulus, while the glucuronide is secreted by the tubules. Accumulation of chloramphenicol in the body may occur if kidney function is poor.

Toxicity. Hypersensitivity reactions, including skin rashes, fever, gastro-intestinal discomfort and soreness of the mouth are relatively trivial side effects of chloramphenicol.

When given to newly-born infants, chloramphenicol was found to induce what is known as the 'grey syndrome' characterised by a failure to eat, hypothermia, muscular flacidity and an ashen grey pallor. Grey syndrome is fatal in about half of all cases but while the cause is not known, the syndrome is associated with the inability of the infant to conjugate and excrete chloramphenicol adequately.

Chloramphenicol also causes suppression of development of some or all blood cells made in the bone marrow. This appears to be a hypersensitivity response, and if all blood-cell production is arrested then it is inevitably fatal. The incidence of bone marrow aplasia is not related to the dose of drug used, and it may occur up to four months after exposure to the drug in about 1 in 40,000 patients.

Reversible dose-dependent depression of the bone marrow leading to reticulocytopoenia and anaemia is not associated with total suppression of bone marrow cells, and is much more common than the disastrous aplasia.

Optic neuritis, characterised as loss of visual acuity, is generally reversible when the drug is stopped. Administration of vitamin B complex is reported to speed recovery of visual acuity following exposure to chloramphenicol.

5.4.4 Fusidate

This compound is well-absorbed when taken orally, and readily penetrates into most tissues except the cerebrospinal fluid. It can also be given intravenously as the diethenolamine salt if necessary.

Excretion is mainly by biliary secretion of the glucuronide conjugate although some metabolites are excreted in the urine. Toxic effects are mild, gastro-intestinal upset and mild skin rashes occuring in some patients.

Interactions. Some antagonism of antibacterial action has been reported when fusidate is tested with penicillins *in vitro*. This is presumably because fusidate will prevent growth of bacteria while penicillins are only effective against growing cells. The significance of this antagonism *in vitro* is difficult to evaluate, but useful complementation of antibacterial action has been reported when fusidate and penicillins are used *in vivo*.

5.4.5 Macrolide Antibiotics

Erythromycin is the most important and widely-used of this group of antibiotics which includes **oleandomycin, spiramycin** and others. The antibacterial spectrum of this group is similar to that of the penicillins.

Erythromycin is not well-absorbed when taken orally, as the erythromycin base is destroyed by acid. **Erythromycin estolate** is frequently used orally and is well-absorbed, but is not bacteriologically active. The estolate has to be hydrolysed mainly in the liver to produce free erythromycin. **Erythromycin glucoheptonate** can be given intravenously if rapid establishment of an effective plasma concentration is required, but intramuscular injections of this preparation are painful.

Erythromycin is distributed in most of the body water compartments and adequate concentrations are found in cells. Penetration into the cerebrospinal fluid is poor however, unless the meninges are inflamed, when adequate therapeutic concentrations may be achieved.

Metabolism and Excretion. A given dose of erythromycin cannot be completely accounted for by excretion via the kidneys and loss in the faeces. Presumably some is metabolised to compounds which have not yet been identified. Biliary secretion probably causes some entero-hepatic circulation, and in man the bile is the major known route of elimination.

Toxicity. Erythromycin base does not have any recognised toxic actions; however, side effects are associated with the use of erythro-mycin estolate. These usually take the form of abdominal pain, fever, enlargement of the liver and decreased biliary secretion (cholestatic hepatitis), and most frequently occur after about two weeks of treat-ment. These effects are believed to be a hypersensitivity reaction to estolate, and whereas recovery is generally satisfactory, recurrence is likely if any estolate is given subsequently. Nausea, vomiting and diarrhoea, and allergic skin reactions also occur in a small percentage of patients.

Oleandomycin is used as the estolate and it appears to have a similar spectrum of action to erythromycin. It differs from erythromycin in not causing cholestatic hepatitis.

FURTHER READING

General Texts

Franklin, T.J. and Snow, G.A. (1975) *Biochemistry of Antimicrobial Action*, 2nd edn (Chapman and Hall, London). Good account of the mechanism of action of common antimicrobials.

Gale, E.F., Cundliffe, E., Reynolds, P.E., Richmond, M.H. and Waring, M.J. (1981) *The Molecular Basis of Antibiotic Action*, 2nd edn (John Wiley, London). Comprehensive treatise on the mechanism of action of antibiotics.

Garrod, L.P., Lambert, H.P., O'Grady, F. and Waterworth, P. (1981) *Antibiotics and Chemotherapy*, 5th edn (Churchill Livingstone, Edinburgh). Covers the microbiological and pharmacological aspects of antibiotics.

Goodman, L.S. and Gilman, A. (1980) *The Pharmacological Basis of Therapeutics*, 6th edn (Macmillan, New York). Comprehensive treatise for the pharmacology of antibiotics.

Hammond, S.M. and Lambert, P.A. (1978) *Antibiotics and Antimicrobial Action* (Edward Arnold, London). Brief general account of antibiotics including the biochemistry of their modes of action.

Pratt, W.B. (1977) *Chemotherapy of Infection* (Oxford University Press, New York). Intermediate-sized book with good accounts of properties, mode of action and pharmacology of a wide range of antibacterials as well as antiviral, antifungal and antiparasitic drugs.

For Specific Chapters

Chapter 2

Brown, G.M. (1962) 'The Biosynthesis of Folic Acid. II Inhibition by Sulphonamides', *Journal of Biological Chemistry, 237*, 536-40

Bushby, S.R.M. and Hitchings, G.H. (1968) 'Trimethoprim, a Sulphonamide Potentiator', *British Journal of Pharmacology and Therapeutics, 33*, 72-90

Woods, D.D. (1962) 'The Biochemical Mode of Action of the Sulphonamide Drugs', *Journal of General Microbiology, 29*, 687-702

Further Reading

Chapter 3

Kahan, J.S., Kahan, F.M., Goegelman, R., Currie, S.A., Jackson, M., Stapley, E.O., Miller, T.W., Miller, A.K., Hendlin, D., Mochales, S., Hernandes, S., Woodruff, H.B. and Birnbaum, J. (1979) 'Thienamycin, a New β-lactam Antibiotic. I. Discovery, Taxonomy, Isolation and Physical Properties', *The Journal of Antibiotics, 32,* 1-12

Kass, E.H. and Evans, D.A. (1979) 'Future Prospects and Past Problems in Antimicrobial Therapy: The Role of Cefoxitin', *Reviews of Infectious Diseases, 1,* 1-239

Richmond, M.H. and Sykes, R.B. (1973) 'The β Lactamases of Gram-Negative Bacteria and their Possible Physiological Roles' in Rose, A.H. and Tempest, D.W. (eds.), *Advances in Microbial Physiology, Vol. 9* (Academic Press, London), 31-88

Shockman, G.D., Daneo-Moore, L., Cornett, J.B. and Mychajlonka, M. (1979) 'Does Penicillin Kill Bacteria?' *Reviews of Infectious Diseases, 1,* 787-96

Woodruff, H.B., Mata, J.M., Hernandez, S., Mochales, S., Rodrigues, A., Stapley, E.O., Wallick, H., Miller, A.K. and Hendlin, D. (1977) 'Fosfomycin: Laboratory Studies', *Chemotherapy, 23* (Supplement 1), 1-22

Chapter 4

Lester, W. (1972) 'Rifampin: A Semisynthetic Derivative of Rifamycin — A Prototype for the Future' in Clifton, C.E., Raffet, S. and Starr, M.P. (eds.), *Annual Review of Microbiology, Vol. 26* (Annual Reviews Inc., Palo Alto, California), 85-102

Riva, S. and Silvestri L.G. (1972) 'Rifamycins: A General View' in Clifton, C.E., Raffel, S. and Starr, M.P. (eds.), *Annual Review of Microbiology, Vol. 26* (Annual Reviewa Inc., Palo Alto, California), 199-225

Chapter 5

Beard, N.S. Jr., Armentrout, S.A. and Weisberger, A.S. (1969) 'Inhibition of Mammalian Protein Synthesis by Antibiotics', *Pharmacological Reviews, 21,* 213-38

Chopra, I. and Howe, T.G.B. (1978) 'Bacterial Resistance to the Tetracyclines' *Microbiological Reviews, 42,* 707-24

Pestka, S. (1971) 'Inhibitors of Ribosome Functions' in Clifton, C.E., Raffel, S. and Starr, M.P. (eds.), *Annual Review of Microbiology,*

Vol 25 (Annual Reviews Inc., Palo Alto, California), 487-562

Weisblum, B. and Davies, J. (1968) 'Antibiotic Inhibitors of the Bacterial Ribosomes', *Bacteriological Reviews, 32*, 493-528

Zierhut, G., Piepersberg, W. and Bock, A. (1979) 'Comparative Analysis of the Effect of Aminoglycosides on Bacterial Protein Synthesis *in vitro*', *European Journal of Biochemistry, 98*, 577-83

INDEX